APPLIED RESEARCH IN FIELD CROP PATHOLOGY FOR INDIANA, 2019

APPLIED RESEARCH IN FIELD CROP PATHOLOGY FOR INDIANA, 2019

DARCY E. P. TELENKO

PURDUE UNIVERSITY PRESS
WEST LAFAYETTE, INDIANA

Cataloging-in-Publication Data on file at the Library Congress.

978-1-62671-356-7 (paperback)

978-1-62671-357-4 (epdf)

Any figures available in color appear that way in the e-book format.

CONTENTS

ACKNOWLEDGEMENTS

This report is a summary of applied field crop pathology research trials conducted in 2019 under the direction of the Purdue Field Crop Pathology program in the Department of Botany and Plant Pathology at Purdue University. The authors wish to thank the Purdue Agronomy Research and Education Center, the Purdue Agricultural Centers, and the many cooperators and contributors who provided the resources needed to support the applied field crop pathology research program in Indiana. Special recognition is extended to Jeffrey Ravellette, Su Shim, and Camila Rocco da Silva for technical skills in managing field trials and data organization and processing and helping prepare this report; Mariama Brown, Tiffanna Ross, Natalia Piñeros, graduate students who assisted with field trial data collection and analysis; Amelia Chaille, Audrey Conrad, Cayla Haupt, Emily Duncan, Doug Keyes, and Kaitlin Waibel, undergraduate student interns who assisted with field trial data collection and scouting; and Dr. Tom Creswell, Dr. John Bonkowski, and Todd Abrahamson with the Purdue Plant Pest Diagnostic Laboratory, who assisted in pathogen surveys and diagnosis. Collectively, the contributions of colleagues, professionals, students, and growers were responsible for a highly successful and productive program to evaluate products and practices for disease management in field crops.

The authors would also like to thank the following for their support in 2019: AMVAC, Bayer Crop Science, BASF, Certis USA, Corteva Agriscience, FFAR—ROAR, FMC Agricultural Solution, Gowan, the Indiana Corn Marketing Council, the Indiana Soybean Alliance, the North Central Soybean Research Program, Pioneer, Purdue University, Sipcam Agro, Syngenta, UPD NA Inc., USDA NIFA Hatch Project #1019253, and USWBSI—NFO.

SUMMARY OF THE 2019 FIELD CROP DISEASE SEASON

CORN

In 2019 most diseases on corn in Indiana remained relatively low across the state with a few exceptions, as listed below. Gray leaf spot, northern corn leaf blight, northern corn leaf spot, and Diplodia streak could be found in pockets. There were also numerous reports of Physoderma brown spot and stalk rot, and there was an increase of many different ear rots caused by insect feeding. Tar spot and southern rust were two diseases that were closely tracked this season.

Tar spot: Tar spot of corn was a concern in 2019 following the localized epidemics experienced in 2018. In 2019 Indiana continued to have localized epidemics, but they were not as widespread as in 2019. The weather is going to be key in determining field risk year to year, as leaf wetness plays an important role in tar spot disease development. The first year of tar spot–directed research has been completed here in Indiana. As a cautionary note, it is important to have multiple years of data for verification, but the initial results do serve as a good starting point for making future management decisions.

The field crop pathology team made a large effort at the end of the season to scout for tar spot across the state. Twenty-five new counties were confirmed with tar spot in 2019, making a total of 65 counties confirmed in Indiana. Out of the 148 fields scouted, 132 were positive for tar spot (89.2%). In addition, incidence and severity were rated (examples of severity are in Figure 1) and used to generate a tar spot index, shown on the map in Figure 1—the darker the county, the greater tar spot index observed in 2019. The map demonstrates how corn produced in northern Indiana is at a higher risk for tar spot versus central and southern Indiana. The map also parallels the ideal weather conditions for tar spot and reports during 2018 and 2019. It is important to document tar spot movement in the state should favorable conditions arise that increase tar spot disease risk across the remainder of the state.

Southern corn rust: Southern corn rust was first confirmed in Indiana on July 25, 2019, and by the end of the season a total of 32 counties were confirmed (Figure 2). Southern rust pustules generally tend to occur on the upper surface of the leaf and produce chlorotic symptoms on the underside of the leaf (Figure 2). These pustules rupture the leaf surface and are orange to tan in color. They are circular to oval in shape. Common rust was also widespread, and both diseases could be present on a leaf and easily mistaken for each other. It is important to send a sample to the Purdue Plant Pest Diagnostic Lab (PPDL) for confirmation if southern rust is suspected. There is an increased risk for yield impact if southern rust is identified early in the season.

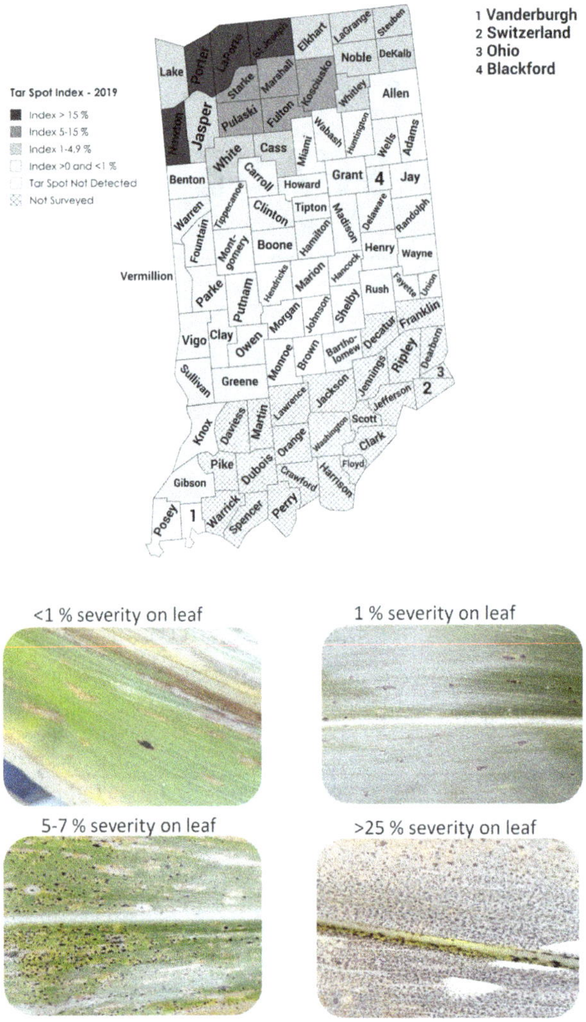

FIGURE 1. 2019 Tar Spot Index for Indiana. (*Top*) The darker the county, the greater the field incidence and severity of tar spot in the fields in which it was found. (*Bottom*) The range of tar spot severity on leaves >25%, 5–7%, and 1% and <1%. Image and photo credit: D. Telenko.

FIGURE 2. (*Left*) Southern corn rust pustules and (*right*) map of confirmed (black) counties that had southern corn rust in Indiana in 2019. Photo credit: D. Telenko, Map source: https://corn.ipmpipe.org/southerncornrust/.

Due to the need to monitor both southern rust and tar spot in Indiana, there will be no charge for southern rust and tar spot samples submitted to the PPDL for diagnostic confirmation again in 2020. This service is made possible through research supported by the Indiana Corn Marketing Council.

SOYBEAN

Diseases in soybeans remained relatively low throughout the season for much of the state in 2019. Our research sites and sentinel plots across the state saw low levels of frogeye leaf spot, Cercospora blight, downy mildew, and Septoria brown spot. There were a few spots of sudden death syndrome and white mold as well. It was a quiet year for diseases in soybean.

WHEAT

Fusarium head blight (FHB), or scab, is one of the most impactful diseases of wheat and the most challenging to prevent. In addition, FHB infection can cause the production of a mycotoxin called deoxynivalenol (DON, or vomitoxin). The conditions in 2019 were extremely conducive to FHB development, and we received numerous reports about issues with FHB and DON contamination in Indiana. Our research sites in both West Lafayette and Vincennes had high levels of FHB develop in our nontreated susceptible variety checks, and initial DON testing was at 7 ppm. Fusarium head blight management requires an integrated approach. This includes selection of varieties with moderate resistance and timely fungicide application at flowering. Other diseases observed in our wheat trials in 2019 included barley yellow dwarf virus, leaf rust, Septoria leaf and glume blotch, and stripe rust.

AGRONOMY CENTER FOR RESEARCH AND EDUCATION (ACRE)

COMPARISON OF FUNGICIDES APPLIED AT TASSEL/SILK (VT/R1) OR MILK (R3) FOR FOLIAR DISEASE IN CORN IN CENTRAL INDIANA, 2019 (COR19-01.ACRE)

C. Rocco da Silva, S. Shim, J. D. Ravellette, and D. E. P. Telenko, Department of Botany and Plant Pathology, Purdue University West Lafayette, Indiana 47907-2054

CORN (*ZEA MAYS* P9998AM)

Gray leaf spot, *Cercospora zeae-maydis*
Northern corn leaf blight, *Exserohilum turcicum*
Common rust, *Puccinia sorghi*
Tar spot, P*hyllachora maydis*

A trial was established at the Purdue Agronomy Center for Research and Education (ACRE) in Tippecanoe County, IN. The experiment was a randomized complete block design with four replications. Plots were 10 feet wide and 30 feet long and consisted of four rows, and the two center rows were used for evaluation. The previous crop was corn. Standard practices for nonirrigated grain corn production in Indiana were followed. Corn hybrid P9998AM was planted in 30-inch row spacing at a rate of 34,000 seeds/acre on June 4 using a John Deere 1700 six-row planter. All fungicide applications were applied at 15 gal/acre and 40 psi using a Lee self-propelled sprayer equipped with a 10-foot boom, fitted with six TJ-VS 8002 nozzles spaced 20 inches apart, at 3.6 mph. Fungicides were applied on August 4 at tassel/silk (VT/R1) and on August 23 at milk (R3) growth stages. Disease ratings were assessed on August 28 and September 23 at the dough (R4) and dent (R5) growth stages, respectively. Disease was rated by visually assessing as a percentage (0–100%) of severity of disease on ear leaves; five plants per plot were rated and averaged. The two center rows of each plot were harvested on October 16, and yields were adjusted to 15.5% moisture. Data were subjected to analysis of variance (SAS 9.4, 2019), and means were compared using Fisher's Least Significant Difference test (LSD; α=0.05).

In 2019, weather conditions were favorable for disease. Gray leaf spot and common rust (CR) were the most prominent diseases in the trial and reached low severity. All fungicide treatments significantly reduced severity of gray leaf spot over the nontreated control at ear leaf on September 23 at both the VT/R1 and

R3 application timings (Table 1). Northern corn leaf blight and tar spot were also found in the trial but at low levels; there was no significant fungicide or timing effect on either disease. All fungicide treatments and timings significantly reduced the severity of common rust over the nontreated control at ear leaves on September 23. There were no significant differences between treatments for percentage of canopy greenness, lodging, ear rot, harvest moisture, test weight, and yield (Table 2).

TABLE 1. ***Effect of Fungicide on Foliar Disease Severity on September 23***

TREATMENT, RATE/ACRE, AND TIMING[z]	GLS[y] %	NCLB[y] %	CR[y] %	TAR SPOT[x] %
Nontreated control	3.8 a	0.1	0.6 a	0.01
Lucento 5.0 fl oz at VT/R1	1.0 b-f	0.2	0.1 bc	0.00
Trivapro 13.7 fl oz at VT/R1	0.9 f	0.0	0.2 bc	0.00
Miravis Neo 13.7 fl oz at VT/R1	0.6 ef	0.0	0.1 bc	0.00
Veltyma 7.0 fl oz at VT/R1	0.3 f	0.0	0.1 bc	0.00
Delaro 8.0 fl oz at VT/R1	1.5 b-e	0.0	0.1 bc	0.00
Quilt Xcel 10.5 fl oz at VT/R1	0.7 def	0.0	0.3 b	0.00
Headline AMP 10.0 fl oz at VT/R1	1.5 b-e	0.1	0.2 bc	0.00
Revytek 8.0 fl oz at VT/R1	0.2 f	0.0	0.1 bc	0.00
Lucento 5.0 fl oz at R3	1.8 bc	0.1	0.0 c	0.00
Trivapro 13.7 fl oz at R3	1.7 bcd	0.1	0.2 bc	0.00
Miravis Neo 13.7 fl oz at R3	2.0 b	0.0	0.1 bc	0.00
Veltyma 7.0 fl oz at R3	0.9 c-f	0.0	0.2 bc	0.00
Delaro 8.0 fl oz at R3	2.0 b	0.0	0.1 bc	0.00
Quilt Xcel 10.5 fl oz at R3	1.6 b-e	0.1	0.1 bc	0.00
Headline AMP 10.0 fl oz at R3	1.9 bc	0.1	0.3 b	0.00
Revytek 8.0 fl oz at R3	1.1 b-f	0.0	0.1 bc	0.00
P-value[w]	*0.0001*	*0.7916*	*0.0090*	*0.4727*

[z] Fungicide treatments were applied on August 4 at the tassel/silk (VT/R1) and August 23 at the milk (R3) growth stages, and all treatments contained a nonionic surfactant (Preference) at a rate of 0.25% v/v.

[y] Disease severity was visually assessed as a percentage (0–100%) of symptomatic leaf area on ear leaves on August 28 and September 23. Five plants were assessed per plot and averaged before analysis. GLS = gray leaf spot, NCLB = northern corn leaf blight, CR = common rust.

[x] Tar spot stroma was visually assessed as a percentage (0–100%) of leaf area on five plants in each plot at ear leaves.

[w] Means followed by the same letter are not significantly different based on Fisher's Least Significant Difference test (LSD; α=0.05).

TABLE 2. ***Effect of Fungicide on Canopy Greenness, Lodging, Ear Rot, and Corn Yield***

TREATMENT, RATE/ACRE, AND TIMING[z]	CANOPY GREEN[y] %	LODGING[x] %	EAR ROT[w] %	HARVEST MOISTURE %	TEST WEIGHT LB/BU	YIELD[v] BU/ACRE
Nontreated control	58.8	0.3	2.2	19.2	55.0	165.7
Lucento 5.0 fl oz at VT/R1	58.8	0.3	1.9	19.2	55.0	161.9
Trivapro 13.7 fl oz at VT/R1	58.8	0.0	3.0	19.2	54.8	163.6
Miravis Neo 13.7 fl oz at VT/R1	53.8	0.0	2.7	18.9	54.5	150.0
Veltyma 7.0 fl oz at VT/R1	60.0	0.0	1.3	19.3	55.0	175.6
Delaro 8.0 fl oz at VT/R1	57.5	0.3	2.3	19.5	54.2	156.6
Quilt Xcel 10.5 fl oz at VT/R1	61.3	0.0	1.4	19.2	54.8	177.7
Headline AMP 10.0 fl oz at VT/R1	58.8	0.0	2.2	19.4	54.5	158.4
Revytek 8.0 fl oz at VT/R1	60.0	0.3	1.8	19.2	54.6	172.2
Lucento 5.0 fl oz at R3	56.3	0.0	1.9	19.5	54.0	157.5
Trivapro 13.7 fl oz at R3	58.8	0.0	2.3	19.1	54.9	163.2
Miravis Neo 13.7 fl oz at R3	58.8	0.0	4.0	19.1	54.5	160.8
Veltyma 7.0 fl oz at R3	57.5	0.3	2.7	19.4	54.4	156.8
Delaro 8.0 fl oz at R3	56.3	0.0	3.9	18.6	54.6	158.0
Quilt Xcel 10.5 fl oz at R3	56.3	0.0	4.4	19.1	54.8	163.1
Headline AMP 10.0 fl oz at R3	55.0	0.3	6.4	19.3	54.8	153.6
Revytek 8.0 fl oz at R3	56.3	0.8	1.6	19.1	54.5	163.8
P-value[u]	*0.3085*	*0.4118*	*0.5087*	*0.61610*	*0.3636*	*0.2056*

[z] Fungicide treatments were applied on August 4 at the tassel/silk (VT/R1) and August 23 at the milk (R3) growth stages, and all treatments contained a nonionic surfactant (Preference) at a rate of 0.25% v/v.

[y] Canopy greenness was visually assessed as a percentage (0–100%) on September 23.

[x] Lodging was calculated as percentage of lodged stalks when pushed from shoulder height to 45° from vertical on September 23.

[w] Ear rot was visually assessed as a percentage (0–100%) from 10 ears per plot on November 8. A mix of ear rot pathogens were identified and included *Fusarium*spp., *Gibberella, Diplodia,* and *Trichoderma* that were associated with significant insect feeding (data not presented).

[v] Yields were adjusted to 15.5% moisture at harvest on November 8.

[u] Means followed by the same letter are not significantly different based on Fisher's Least Significant Difference test (LSD; α=0.05).

FUNGICIDE EVALUATION FOR TAR SPOT AND OTHER FOLIAR DISEASES IN CORN IN CENTRAL INDIANA, 2019 (COR19-02.ACRE)

T. Ross, S. Shim, J. D. Ravellette, and D. E. P. Telenko, Department of Botany and Plant Pathology, Purdue University West Lafayette, Indiana 47907-2054

CORN (*ZEA MAYS* P9998AM)

Gray leaf spot, *Cercospora zeae-maydis*

A trial was established at the Purdue Agronomy Center for Research and Education (ACRE) in Tippecanoe County, IN. The experiment was a randomized complete block design with four replications. Plots were 10 feet wide and 30 feet long and consisted of four rows, and the two center rows were used for evaluation. The previous crop was corn. Standard practices for nonirrigated grain corn production in Indiana were followed. Corn hybrid P9998AM was planted in 30-inch row spacing at a rate of 34,000 seeds/acre on June 4 using a John Deere 1700 six row planter. All fungicide applications were applied at 15 gal/acre and 40 psi using a Lee self-propelled sprayer equipped with a 10-foot boom, fitted with six TJ-VS 8002 nozzles spaced 20 inches apart, at 3.6 mph. Fungicides were applied on August 4 at tassel/silk (VT/R1) growth stage. Disease ratings were assessed on August 28 and September 23 at the early dough (R4) and dent (R5) growth stages, respectively. Disease severity was visually assessed as a percentage (0=100%) of symptomatic leaf area on ear leaf; five leaves were assessed per plot and averaged before analysis. The two center rows of each plot were harvested on October 16, and yields were adjusted to 15.5% moisture. Data were subjected to analysis of variance (SAS 9.4, 2019), and means were compared using Fisher's Least Significant Difference test (LSD; α=0.05).

Gray leaf spot was the most prominent disease in the trial and reached low severity. All fungicide treatments reduced severity of gray leaf spot over the nontreated control on September 23 (Table 3). All treatments increased the percent canopy greenness over the nontreated except Headline AMP. Veltyma treatment resulted in the greenest canopy but was not different from Trivapro, Delaro, Quilt Xcel, Headline SC, or Proline. There was no significant difference between treatments for lodging, harvest moisture, test weight, and yield.

TABLE 3. ***Effect of Fungicide on Foliar Disease Severity and Corn Yield***

TREATMENT AND RATE/ACRE[z]	GLS[y] % EL	CANOPY GREEN[x] %	LODGING[w] %	HARVEST MOISTURE %	TEST WEIGHT LB/BU	YIELD[v] BU/ACRE
Nontreated control	4.8 a	53.8 d	0.0	20.0	54.7	171.2
Trivapro 2.21 SE 13.7 fl oz	0.0 b	62.5 ab	0.5	20.7	54.8	173.0
Miravis Neo 2.5 SE 13.7 fl oz	0.0 b	60.0 bc	0.3	20.4	54.6	159.4
Delaro 325 SC 12.0 fl oz	0.0 b	61.3 abc	0.0	20.5	54.1	164.9
Headline AMP 1.68 SC 14.4 fl oz	0.0 b	56.3 cd	0.0	19.5	55.1	182.2
Topguard EQ 7.0 fl oz	0.0 b	60.0 bc	0.3	20.9	54.1	177.1
Quilt Xcel 2.2 SE 14.0 fl oz	0.0 b	61.3 abc	0.0	19.9	55.0	179.1
Veltyma 3.34 SC 7.0 fl oz	0.0 b	66.3 a	0.0	20.5	54.5	181.1
Aproach Prima 2.34 SC 6.8 fl oz	0.0 b	60.0 bc	0.0	20.7	54.2	166.0
Headline 2.09 SC 12.0 fl oz	0.0 b	61.3 abc	0.0	20.7	54.0	171.7
Proline 480 SC 5.7 fl oz	0.0 b	61.3 abc	0.0	20.4	54.4	174.8
P-value[u]	*0.0001*	*0.0123*	*0.1162*	*0.5329*	*0.3175*	*0.6710*

[z] Fungicide treatments were applied on August 4 tassel/silk (VT/R1) growth stage, and all treatments contained a nonionic surfactant (Preference) at a rate of 0.25% v/v.

[y] Disease severity was visually assessed as a percentage (0–100%) of symptomatic leaf area on ear leaf on September 23. Five plants were assessed per plot and averaged before analysis. GLS = gray leaf spot.

[x] Canopy green was visually assessed as a percentage (0–100%) on September 23.

[w] Lodging = percentage of lodged stalks when pushed from shoulder height to 45° from vertical on September 23.

[v] Yields were adjusted to 15.5% moisture at harvest on October 16.

[u] Means followed by the same letter are not significantly different based on Fisher's Least Significant Difference test (LSD; α=0.05).

UNIFORM FUNGICIDE TIMING AND TAR SPOT MODEL VALIDATION IN CORN IN CENTRAL INDIANA, 2019 (COR19-04.ACRE)

S. Shim, J. D. Ravellette, and D. E. P. Telenko, Department of Botany and Plant Pathology, Purdue University
West Lafayette, Indiana 47907-2054

CORN (*ZEA MAYS* P9998AM)

Gray leaf spot, *Cercospora zeae-maydis*

A trial was established at the Purdue Agronomy Center for Research and Education (ACRE) in Tippecanoe County, IN. The experiment was a randomized complete block design with four replications. Plots were 10 feet wide and 30 feet long and consisted of four rows, and the two center rows were used for evaluation. The previous crop was corn. Standard practices for nonirrigated grain corn production in Indiana were followed. Corn hybrid P9998AM was planted in 30-inch row spacing at a rate of 34,000 seeds/acre on June 4. All fungicide applications were applied at 15 gal/acre and 40 psi using a Lee self-propelled sprayer equipped with a 10-foot boom, fitted with six TJ-VS 8002 nozzles spaced 20 inches apart, at 3.6 mph. The fungicide Trivapro 2.21 SE was applied on July 5 at V6 growth stage, July 11 at V8 growth stage, July 17 at V10 growth stage, August 4 at the tassel/silk (VT/R1) growth stage, and August 16 at blister (R2) growth stage. Disease ratings were assessed on August 29 at the early dough (R4) growth stage and on October 16 at the maturity (R6). Disease severity was visually assessed as a percentage (0=100%) of upper or lower canopy on August 29. The two center rows of each plot were harvested on October 16, and yields were adjusted to 15.5% moisture. Data were subjected to analysis of variance (SAS 9.4, 2019), and means were compared using Fisher's Least Significant Difference test (LSD; α=0.05).

In 2019, gray leaf spot was the most prominent disease in the trial and reached low severity. All fungicide treatments and timings reduced the severity of gray leaf spot in upper and lower canopies as compared to the nontreated control on August 29 (Table 4). There was no significant difference between treatments for percentage of canopy greenness, moisture, test weight, and yield.

TABLE 4. ***Effect of Fungicide and Timing on Foliar Disease Severity and Corn Yield***

TREATMENT, RATE/ACRE, TIMING[z]	GLS[y] % LOWER CANOPY	GLS[y] % UPPER CANOPY	CANOPY GREEN[x] %	HARVEST MOISTURE %	TEST WEIGHT LB/BU	YIELD[w] BU/ACRE
Nontreated control	11.3 a	2.0 a	55.0	20.5	55.2	177.0
Trivapro 2.21 SE 13.7 fl oz at V6	5.0 b	1.0 bc	57.5	20.1	54.8	182.4
Trivapro 2.21 SE 13.7 fl oz at V8	5.0 b	0.5 c	60.0	19.9	54.9	178.6
Trivapro 2.21 SE 13.7 fl oz at V10	5.0 b	0.5 c	53.8	19.9	54.9	173.4
Trivapro 2.21 SE 13.7 fl oz at VT	5.0 b	0.8 bc	60.0	20.0	54.6	182.5
Trivapro 2.21 SE 13.7 fl oz at R2	6.3 b	0.8 bc	56.3	19.3	55.0	181.6
Trivapro 2.21 SE 13.7 fl oz V6 fb VT	5.0 b	0.5 c	60.0	20.2	54.4	186.3
Trivapro 2.21 SE 13.7 fl oz at tar spot model	6.3 b	1.3 b	51.3	19.8	55.2	178.0
P-value[v]	*0.0001*	*0.0005*	*0.2990*	*0.8034*	*0.1932*	*0.6381*

[z] Fungicide treatment Trivapro 2.21 SE was applied on July 5 at V6 growth stage, July 11 at V8 growth stage, July 17 at V10 growth stage, August 4 at the tassel/silk (VT/R1) growth stage, and August 16 at R2 (blister) growth stage. All treatments did not contain a nonionic surfactant. Tar spot model based application was not applied. fb = followed by.

[y] Disease severity was visually assessed as a percentage (0=100%) of upper or lower canopy on August 29. GLS = gray leaf spot.

[x] Canopy green was visually assessed as a percentage (0–100%) on September 23.

[w] Yields were adjusted to 15.5% moisture at harvest on October 16.

[v] Means followed by the same letter are not significantly different based on Fisher's Least Significant Difference test (LSD; α=0.05).

EVALUATION OF FUNGICIDES FOR FOLIAR DISEASE IN CORN IN CENTRAL INDIANA, 2019 (COR19-19.ACRE)

S. Shim, J. D. Ravellette, and D. E. P. Telenko, Department of Botany and Plant Pathology, Purdue University
West Lafayette, Indiana 47907-2054

CORN (*ZEA MAYS* P9998AM)

Gray leaf spot, *Cercospora zeae-maydis*
Northern corn leaf blight, *Exserohilum turcicum*
Southern rust, *Puccinia polysora*

A trial was established at the Purdue Agronomy Center for Research and Education (ACRE) in Tippecanoe County, IN. The experiment was a randomized complete block design with four replications. Plots were 10 feet wide and 30 feet long and consisted of four rows, and the two center rows were used for evaluation. The previous crop was corn. Standard practices for nonirrigated grain corn production in Indiana were followed. Corn hybrid P9998AM was planted in 30-inch row spacing at a rate of 34,000 seeds/acre on June 4. All fungicide applications were applied at 15 gal/acre and 40 psi using a Lee self-propelled sprayer equipped with a 10-foot boom, fitted with six TJ-VS 8002 nozzles spaced 20 inches apart, at 3.6 mph. Fungicides were applied on July 1 at the V5 growth stage and on August 4 at the tassel/silk (VT/R1) growth stage. Disease ratings were assessed on August 28 and September 23 at the dough (R4) and dent (R5) growth stages, respectively. Disease severity was visually assessed as percentage (0–100%) of symptomatic leaf area on ear leaf; five plants were assessed per plot and averaged before analysis. The two center rows of each plot were harvested on October 16, and yields were adjusted to 15.5% moisture. Data were subjected to analysis of variance (SAS 9.4, 2019), and means were compared using Fisher's Least Significant Difference test (LSD; α=0.05).

In 2019, gray leaf spot, northern corn leaf blight, and southern rust were the most prominent diseases in the trial and reached low severity. All fungicides significantly reduced gray leaf spot severity over the nontreated control on September 23 (Table 5). The Brixen application at V5 had significantly higher gray leaf spot than the other fungicides and timings. No differences between treatments and nontreated control were detected for northern corn leaf blight. Southern rust was significantly reduced by all fungicide applications and timings over the nontreated control. Veltyma, Trivapro, Fortix, and two applications of Brixen had the lowest southern rust but were only significantly different than Brixen applied at V5. Brixen applied at VT/R1, Fortix, USF0411, Trivapro, and Quilt Xcel significantly increased canopy greenness over the nontreated control on September 23 (Table 6). There was no significant difference between treatments for lodging, moisture, test weight and corn yield.

TABLE 5. ***Effect of Fungicide on Foliar Disease Severity***

TREATMENT, RATE/ACRE, AND TIMING[z]	GLS[y] %	NCLB[y] %	SR[y] %
Nontreated control	4.4 a	0.3	1.00 a
Brixen 15.0 fl oz at V5	2.9 b	0.0	0.33 b
Brixen 15.0 fl oz at V5 fb 13.0 fl oz at VT/R1	0.3 c	0.0	0.01 c
Brixen 3.0 fl oz at VT/R1	0.5 c	0.2	0.06 bc
Fortix 5.0 fl oz at VT/R1	0.6 c	0.0	0.02 c
Delaro Complete 458 SC 8.0 fl oz at VT/R1	0.7 c	0.8	0.11 bc
Miravis Neo 2.5 SE 13.7 fl oz at VT/R1	0.4 c	0.1	0.11 bc
Trivapro 2.21 SE 13.7 fl oz at VT/R1	0.6 c	0.0	0.00 c
Veltyma 3.34 SC 7.0 fl oz at VT/R1	0.3 c	0.0	0.00 c
Quilt Xcel 2.2 SE 10.5 fl oz at VT/R1	0.8 c	0.1	0.16 bc
Headline AMP 1.68 SC 10.0 fl oz at VT/R1	1.2 c	0.1	0.10 bc
P-value[x]	*0.0001*	*0.3508*	*0.0001*

[z] Fungicide treatments were applied on July 1 at the V5 growth stage and on August 4 at the tassel/silk (VT/R1) growth stage. All treatments except Delaro Complete 458 SC contained a nonionic surfactant (Preference) at a rate of 0.25% v/v. The treatment with Delaro Complete 458 SC contained Induce at a rate of 0.12% v/v. fb = followed by.

[y] Disease severity was visually assessed as a percentage (0–100%) of symptomatic leaf area on ear leaf on September 23. Five plants were assessed per plot and averaged before analysis. GLS = gray leaf spot, NCLB = northern corn leaf blight, SR = southern rust.

[x] Means followed by the same letter are not significantly different based on Fisher's Least Significant Difference test (LSD; α=0.05).

TABLE 6. ***Effect of Fungicide on Canopy Greenness, Lodging, and Corn Yield***

TREATMENT, RATE/ACRE, AND TIMING[z]	CANOPY GREEN[y] %	LODGING[x] %	HARVEST MOISTURE %	TEST WEIGHT LB/BU	YIELD[w] BU/ACRE
Nontreated control	52.5 c	2.5	19.3	55.3	170.1
Brixen 15.0 fl oz at V5	57.5 abc	0.0	19.7	54.7	172.3
Brixen 15.0 fl oz at V5 fb 13.0 fl oz at VT/R1	57.5 abc	2.5	19.8	54.7	176.4
Brixen 3.0 fl oz at VT/R1	61.3 a	0.0	20.2	54.6	174.9
Fortix 5.0 fl oz at VT/R1	58.8 ab	0.0	19.5	54.6	182.6
Delaro Complete 458 SC 8.0 fl oz at VT/R1	58.8 ab	0.0	19.4	55.1	185.7
Miravis Neo 2.5 SE 13.7 fl oz at VT/R1	57.5 abc	5.0	19.6	54.2	175.7
Trivapro 2.21 SE 13.7 fl oz at VT/R1	58.8 ab	0.0	19.6	55.1	171.8
Veltyma 3.34 SC 7.0 fl oz at VT/R1	55.0 bc	0.0	20.0	55.0	178.8
Quilt Xcel 2.2 SE 10.5 fl oz at VT/R1	58.8 ab	0.0	20.1	54.8	183.3
Headline AMP 1.68 SC 10.0 fl oz at VT/R1	57.5 abc	0.0	19.7	54.5	175.0
P-value[v]	*0.1372*	*0.1989*	*0.9412*	*0.1660*	*0.6702*

[z] Fungicide treatments were applied on July 1 at the V5 growth stage and on August 4 at the tassel/silk (VT/R1) growth stage. All treatments except Delaro Complete 458 SC contained a nonionic surfactant (Preference) at a rate of 0.25% v/v. The treatment with Delaro Complete 458 SC contained Induce at a rate of 0.12% v/v. fb = followed by.

[y] Canopy greenness was visually assessed as a percentage (0–100%) of canopy on September 23.

[x] Lodging is the percentage of lodged stalks when pushed from shoulder height to 45° from vertical on September 23.

[w] Yields were adjusted to 15.5% moisture at harvest on October 16.

[v] Means followed by the same letter are not significantly different based on Fisher's Least Significant Difference test (LSD; α=0.05).

EVALUATION OF FUNGICIDE TIMING AND EFFICACY FOR FOLIAR DISEASES IN SOYBEAN IN CENTRAL INDIANA, 2019 (SOY19-01.ACRE)

C. Rocca da Silva, S. Shim, J. D. Ravellette, and D. E. P. Telenko, Department of Botany and Plant Pathology, Purdue University West Lafayette, Indiana 47907-2054

SOYBEAN (*GLYCINE MAX* P34A13X)

Frogeye leaf spot, *Cercospora sojina*
Septoria brown spot, *Septoria glycines*
Cercospora leaf blight, *Cercospora kikuchii*

A trial was established at the Purdue Agronomy Center for Research and Education (ACRE) in Tippecanoe County, IN. The experiment was a randomized complete block design with four replications. Plots were 10 feet wide and 30 feet long and consisted of four rows, and the two center rows were used for evaluation. The previous crop was corn. Standard practices for nonirrigated soybean production in Indiana were followed. Soybean variety P34A13X was planted in 30-inch row spacing at a rate of 140,000 seeds/acre on June 4. All fungicide applications were applied at 15 gal/acre and 40 psi using a Lee self-propelled sprayer equipped with a 10-foot boom, fitted with six TJ-VS 8002 nozzles spaced 20 inches apart, at 3.6 mph. Fungicides were applied on August 5 at the beginning pod (R3) growth stage. Diseases were assessed on September 20 at the full seed (R6) growth stage. Frogeye leaf spot, Cercospora leaf blight, and Septoria brown spot were rated for disease severity by visually assessing the percentage of symptomatic leaf area in the canopy of each plot. The two center rows were harvested on October 17, and yields were adjusted to 13.0% moisture. Data were subjected to analysis of variance (SAS 9.4, 2019), and means were compared using Fisher's Least Significant Difference test (LSD; α=0.05).

In 2019, frogeye leaf spot, Septoria brown spot, and Cercospora light blight were the most prominent diseases in the trial and reached low severity. The fungicide treatments of Quadris Top, Lucento, Priaxor Xemium, Headline AMP, Veltyma, Revytek, Trivapro, and Preemptor significantly reduced the disease severity of Septoria brown spot over the nontreated control on September 20 (Table 7). No significant treatment effects were detected for frogeye leaf spot and Cercospora leaf blight on September 20, harvest moisture, test weight, and soybean yield.

TABLE 7. ***Effect of Fungicide on Foliar Disease Severity and Soybean Yield***

TREATMENT AND RATE/ACRE[z]	FLS[y] %	SBS[y] %	CLB[y] %	HARVEST MOISTURE %	TEST WEIGHT LB/BU	YIELD BU/ACRE[x]
Nontreated control	0.8	0.8 a	0.4	12.8	55.9	51.1
Preemptor SC 5.0 fl oz	0.8	0.3 bc	0.1	12.7	55.8	49.7
Topguard EQ 5.0 fl oz	0.5	0.4 abc	0.3	12.6	55.3	49.6
Quadris Top SBX 3.76 SC 7.0 fl oz	0.5	0.0 c	0.3	12.8	55.9	50.6
Lucento 4.17 SC 5.0 fl oz	0.1	0.0 c	0.3	12.8	55.7	50.0
Miravis Top 1.67 SC 13.7 fl oz	0.4	0.3 bc	0.6	12.8	56.0	51.2
Priaxor Xemium 4.0 fl oz	0.6	0.0 c	0.1	12.9	55.7	53.1
Trivapro 2.21 SE 13.0 fl oz	0.6	0.3 bc	0.3	12.7	55.4	51.7
Delaro 325 SC 8.0 fl oz	0.3	0.5 ab	0.3	12.7	55.9	51.1
Headline AMP 1.68 SC 10.0 fl oz	0.8	0.1 c	0.3	12.9	55.7	52.0
Veltyma 3.34 SC 7.0 fl oz	0.6	0.1 c	0.5	13.0	55.3	52.4
Revytek 3.33 LC 8.0 fl oz	0.3	0.1 c	0.1	12.7	55.9	52.0
P-value[w]	*0.6862*	*0.0090*	*0.8782*	*0.3977*	*0.4031*	*0.7448*

[z] Fungicide treatments were applied on August 5 at the beginning pod (R3) growth stage, and all treatments contained a nonionic surfactant (Preference) at a rate of 0.25% v/v.

[y] Disease severity was visually assessed as a percentage (0–100%) of symptomatic leaf area on September 20. FLS = frogeye leaf spot, SBS = Septoria brown spot, CLB = Cercospora leaf blight.

[x] Yields were adjusted to 13.0% moisture at harvest on October 17.

[w] Means followed by the same letter are not significantly different based on Fisher's Least Significant Difference test (LSD; α=0.05).

EVALUATION OF FUNGICIDES FOR FOLIAR DISEASES ON SOYBEAN IN CENTRAL INDIANA: TRIAL 1, 2019 (SOY19-16.ACRE)

C. Rocco da Silva, S. Shim, J. D. Ravellette, and D. E. P. Telenko, Department of Botany and Plant Pathology, Purdue University West Lafayette, Indiana 47907-2054

SOYBEAN (*GLYCINE MAX* P34A13X)

Frogeye leaf spot, *Cercospora sojina*
Cercospora leaf blight, *Cercospora kikuchii*

A trial was established at the Purdue Agronomy Center for Research and Education (ACRE) in Tippecanoe County, IN. The experiment was a randomized complete block design with four replications. Plots were 10 feet wide and 30 feet long and consisted of four rows, and the two center rows were used for evaluation. The previous crop was corn. Standard practices for nonirrigated soybean production in Indiana were followed. Soybean variety P34A13X was planted in 30-inch row spacing at a rate of 140,000 seeds/acre on June 4. All fungicide applications were applied at 15 gal/acre and 40 psi using a Lee self-propelled sprayer equipped with a 10-foot boom, fitted with six TJ-VS 8002 nozzles spaced 20 inches apart, at 3.6 mph. Fungicides were applied on August 5 at the beginning pod (R3) growth stage. Disease ratings were assessed on September 20 at the full seed (R6) growth stage. Frogeye leaf spot and Cercospora leaf blight were rated for disease severity by visually assessing the percentage of symptomatic leaf area in the upper and lower canopies. The two center rows were harvested on October 17, and yields were adjusted to 13.0% moisture. Data were subjected to analysis of variance (SAS 9.4, 2019), and means were compared using Fisher's Least Significant Difference test (LSD; α=0.05).

In 2019, frogeye leaf spot and Cercospora leaf blight were the most prominent diseases in the trial and reached a low severity. There were no significant treatment effects for the disease severity of frogeye leaf spot in the upper or lower canopies, Cercospora leaf blight in the upper canopy, harvest moisture, test weight, and yield (Table 8).

TABLE 8. *Effect of Fungicide on Foliar Disease Severity and Soybean Yield*

TREATMENT AND RATE/ACRE[z]	FLS[y] % UPPER CANOPY	FLS[y] % LOWER CANOPY	CLB[y] % UPPER CANOPY	HARVEST MOISTURE %	TEST WEIGHT LB/BU	YIELD[x] BU/ACRE
Nontreated control	0.8	0.8	0.5	11.4	55.0	51.1
Delaro 325 SC 8.0 fl oz	0.9	0.2	0.3	11.4	55.4	54.8
Miravis Neo 2.5 SE 13.7 fl oz	0.4	0.1	0.3	11.4	55.6	53.4
Lucento 4.17 SC 5.0 fl oz	0.3	0.1	0.5	11.4	55.5	53.6
Priaxor Xemium 4.0 fl oz	0.7	0.8	0.3	11.5	55.5	52.9
Veltyma 3.34 SC 7.0 fl oz	0.1	0.3	0.0	11.5	55.4	51.3
Revytek 3.33 LC 8.0 fl oz	0.9	0.3	0.3	11.4	55.4	55.5
Trivapro 2.21 SE 13.7 fl oz	0.7	0.3	0.3	11.3	55.8	52.8
P-value[w]	*0.6383*	*0.0677*	*0.7311*	*0.4732*	*0.1084*	*0.7378*

[z] Fungicide treatments were applied on August 5 at the beginning pod (R3) growth stage, and all treatments contained a nonionic surfactant (Preference) at a rate of 0.25% v/v.

[y] Foliar disease severity was rated on a scale of 0–100% within the upper or lower canopy on September 20. FLS = frogeye leaf spot, CLB = Cercospora leaf blight.

[x] Yields were adjusted to 13.0% moisture at harvest on October 17.

[w] Means followed by the same letter are not significantly different based on Fisher's Least Significant Difference test (LSD; α=0.05).

EVALUATION OF FUNGICIDES FOR FOLIAR DISEASES ON SOYBEAN IN CENTRAL INDIANA: TRIAL 2, 2019 (SOY19-17.ACRE)

S. Shim, J. D. Ravellette, and D. E. P. Telenko, Department of Botany and Plant Pathology, Purdue University
West Lafayette, Indiana 47907-2054

SOYBEAN (*GLYCINE MAX* P34A13X)

Frogeye leaf spot, *Cercospora sojina*
Septoria brown spot, *Septoria glycines*

A trial was established at the Purdue Agronomy Center for Research and Education (ACRE) in Tippecanoe County, IN. The experiment was a randomized complete block design with four replications. Plots were 10 feet wide and 30 feet long and consisted of four rows, and the two center rows were used for evaluation. The previous crop was corn. Standard practices for nonirrigated soybean production in Indiana were followed. Soybean variety P34A13X was planted in 30-inch row spacing at a rate of 140,000 seeds/acre on June 4. All fungicide applications were applied at 15 gal/acre and 40 psi using a Lee self-propelled sprayer equipped with a 10-foot boom, fitted with six TJ-VS 8002 nozzles spaced 20 inches apart, at 3.6 mph. Fungicides were applied on August 5 at the beginning pod (R3) growth stage. Disease ratings were assessed on August 28 at the beginning seed (R5) and on September 20 at the full seed (R6) growth stages. Frogeye leaf spot and Septoria brown spot were rated by visually assessing the percentage of symptomatic leaf area in the upper and lower canopies, respectively. The two center rows were harvested on October 17, and yields were adjusted to 13.0% moisture. Data were subjected to analysis of variance (SAS 9.4, 2019), and means were compared using Fisher's Least Significant Difference test (LSD; α=0.05).

In 2019, frogeye leaf spot and Septoria brown spot were the most prominent diseases in the trial and reached low severity. All fungicide treatments reduced the disease severity of frogeye leaf spot on August 28 and on September 20 and of Septoria brown spot on September 20 (Table 9). No significant treatment effects were detected for soybean harvest moisture, test weight, and yield.

TABLE 9. ***Effect of Fungicide on Foliar Disease Severity and Soybean Yield***

TREATMENT AND RATE/ACRE[z]	FLS[y] % AUGUST 28	FLS[y] % UPPER CANOPY SEPTEMBER 20	SBS[y] % LOWER CANOPY SEPTEMBER 20	HARVEST MOISTURE %	TEST WEIGHT LB/BU	YIELD[x] BU/ACRE
Nontreated control	0.1 a	2.0 a	1.3 a	11.8	55.7	46.2
Miravis Top 1.67 SC 13.7 fl oz	0.0 b	0.3 b	0.0 b	12.0	55.7	46.0
Miravis Neo 2.5 SE 13.7 fl oz	0.0 b	0.1 b	0.0 b	11.9	55.7	44.4
Delaro Complete 458 SC 8.0 fl oz	0.0 b	0.3 b	0.1 b	11.8	55.7	47.9
Delaro 325 SC 8.0 fl oz	0.0 b	0.3 b	0.0 b	12.0	56.3	50.8
Lucento 4.17 SC 5.0 fl oz	0.0 b	0.1 b	0.1 b	11.9	55.9	45.4
Aproach Prima 2.34 SC 6.8 fl oz	0.0 b	0.6 b	0.3 b	11.8	55.8	51.4
Priaxor Xemium 4.0 fl oz	0.0 b	0.4 b	0.1 b	12.0	55.8	46.4
Headline AMP 1.68 SC 10.0 fl oz	0.0 b	0.3 b	0.3 b	11.9	55.9	48.3
Veltyma 3.34 SC 7.0 fl oz	0.0 b	0.1 b	0.3 b	11.9	55.7	46.0
P-value[w]	*0.0130*	*0.0380*	*0.0024*	*0.8707*	*0.6725*	*0.7687*

[z] Fungicide treatments were applied on August 5 at the beginning pod (R3) growth stage, and all treatments contained a nonionic surfactant (Preference) at a rate of 0.25% v/v.

[y] Foliar disease severity was rated on a scale of 0–100% for the whole plot on August 28 and of upper and lower canopies on September 20. FLS = frogeye leaf spot, SBS = Septoria brown spot.

[x] Yields were adjusted to 13.0% moisture at harvest on October 17.

[w] Means followed by the same letter are not significantly different based on Fisher's Least Significant Difference test (LSD; α=0.05).

FUNGICIDE COMPARISON FOR FOLIAR DISEASES IN SOYBEAN IN CENTRAL INDIANA, 2019 (SOY19-19.ACRE)

S. Shim, J. D. Ravellette, and D. E. P. Telenko, Department of Botany and Plant Pathology, Purdue University West Lafayette, Indiana 47907-2054

SOYBEAN (*GLYCINE MAX* P34A13X)

Frogeye leaf spot, *Cercospora sojina*
Septoria brown spot, *Septoria glycines*
Cercospora leaf blight, *Cercospora kikuchii*

A trial was established at the Purdue Agronomy Center for Research and Education (ACRE) in Tippecanoe County, IN. The experiment was a randomized complete block design with four replications. Plots were 10 feet wide and 30 feet long and consisted of four rows, and the two center rows were used for evaluation. The previous crop was corn. Standard practices for nonirrigated soybean production in Indiana were followed. Soybean variety P34A13X was planted in 30-inch row spacing at a rate of 140,000 seeds/acre on June 4. All fungicide applications were applied at 15 gal/acre and 40 psi using a Lee self-propelled sprayer equipped with a 10-foot boom, fitted with six TJ-VS 8002 nozzles spaced 20 inches apart, at 3.6 mph. Fungicides were applied on August 5 at the beginning pod (R3) growth stage. Disease ratings were assessed on August 29 at the beginning seed (R4) and on September 20 at the full seed (R6) growth stages. Frogeye leaf spot, Cercospora leaf blight, and Septoria brown spot foliar disease severity were rated on a scale of 0–100% within the upper or lower canopies. The two center rows were harvested on October 17, and yields were adjusted to 13.0% moisture. Data were subjected to analysis of variance (SAS 9.4, 2019), and means were compared using Fisher's Least Significant Difference test (LSD; α=0.05).

In 2019, frogeye leaf spot, Cercospora leaf blight, and Septoria brown spot were the most prominent diseases in the trial and reached low severity. All fungicide treatments reduced Septoria brown spot on September 20 (Table 10). No significant treatment effects were detected for frogeye leaf spot or Cercospora leaf blight severity on September 20, harvest moisture, test weight, and soybean yield.

TABLE 10. ***Effect of Fungicide on Foliar Disease Severity and Soybean Yield***

TREATMENT AND RATE/ACRE[z]	FLS[y] %	SBS[y] %	CLB[y] %	HARVEST MOISTURE %	TEST WEIGHT LB/BU	YIELD[x] BU/ACRE
Nontreated control	1.3	2.5 a	0.3	11.7	55.9	50.0
Topguard EQ 5.0 fl oz	0.4	0.2 b	0.0	11.7	55.6	55.5
Lucento 4.17 SC 5.0 fl oz	0.1	0.1 b	0.0	11.6	55.6	52.7
Lucento 4.17 SC 5.0 fl oz + Hero 5.0 fl oz	0.3	0.0 b	0.0	11.8	55.8	57.2
VJR90-R002 7.0 fl oz	0.3	0.6 b	0.3	11.9	55.5	51.2
VJR90-R002 9.0 fl oz	0.3	0.3 b	0.0	11.8	55.8	49.6
Miravis Top 1.67 SC 13.7 fl oz	0.5	0.5 b	0.1	11.9	55.8	46.4
Priaxor Xemium 4.0 fl oz + Tilt 4.0 fl oz	1.3	0.5 b	0.3	11.7	56.1	52.3
Delaro 325 SC 8.0 fl oz	0.3	0.2 b	0.0	11.7	55.8	50.0
Quadris Top SBX 3.76 SC 7.0 fl oz	0.7	0.3 b	0.0	11.8	55.8	50.6
P-value[w]	*0.4588*	*0.0003*	*0.4635*	*0.6524*	*0.1883*	*0.1973*

[z] Fungicide treatments were applied on August 5 at the beginning pod (R3) growth stage, and all treatments contained a nonionic surfactant (Preference) at a rate of 0.25% v/v.

[y] Foliar disease severity was rated on a scale of 0–100% within the canopy on September 20. FLS = frogeye leaf spot, SBS = Septoria brown spot, CLB = Cercospora leaf blight.

[x] Yields were adjusted to 13.0% moisture at harvest on October 17.

[w] Means followed by the same letter are not significantly different based on Fisher's Least Significant Difference test (LSD; α=0.05).

EVALUATION OF FUNGICIDE TIMING AND EFFICACY FOR FOLIAR DISEASES IN SOYBEAN IN CENTRAL INDIANA, 2019 (SOY19-20.ACRE)

S. Shim, J. D. Ravellette, and D. E. P. Telenko, Department of Botany and Plant Pathology, Purdue University
West Lafayette, Indiana 47907-2054

SOYBEAN (*GLYCINE MAX* P34A13X)

Frogeye leaf spot, *Cercospora sojina*
Septoria brown spot, *Septoria glycines*
Cercospora leaf blight, *Cercospora kikuchii*

A trial was established at the Purdue Agronomy Center for Research and Education (ACRE) in Tippecanoe County, IN. The experiment was a randomized complete block design with four replications. Plots were 10 feet wide and 30 feet long and consisted of four rows, and the two center rows were used for evaluation. The previous crop was corn. Standard practices for nonirrigated soybean production in Indiana were followed. Soybean variety P34A13X was planted in 30-inch row spacing at a rate of 140,000 seeds/acre on June 4. All fungicide applications were applied at 15 gal/acre and 40 psi using a Lee self-propelled sprayer equipped with a 10-foot boom, fitted with six TJ-VS 8002 nozzles spaced 20 inches apart, at 3.6 mph. Fungicides were applied on August 5 at the beginning pod (R3) and on August 16 at the beginning seed (R5) growth stages. Disease ratings were assessed on September 20 at the full seed (R6) growth stage. Frogeye leaf spot, Cercospora leaf blight, and Septoria brown spot were rated by visually assessing the percentage of symptomatic leaf area in the upper and lower canopies. The two center rows were harvested on October 17, and yields were adjusted to 13.0% moisture. Data were subjected to analysis of variance (SAS 9.4, 2019), and means were compared using Fisher's Least Significant Difference test (LSD; α=0.05).

In 2019, frogeye leaf spot, Cercospora leaf blight, and Septoria brown spot were the most prominent diseases in the trial and reached low severity. All fungicide applications and timings reduced frogeye leaf spot in the upper canopy over the nontreated control except Priaxor Xemium at R3 (Table 11). No significant treatment effects were found for disease severity of Septoria brown spot, Cercospora leaf blight, harvest moisture, test weight, and soybean yield.

TABLE 11. ***Effect of Fungicide on Foliar Disease Severity***

TREATMENT, RATE/ACRE, AND TIMING[z]	FLS[y] %	SBS[y] %	CLB[y] %	HARVEST MOISTURE %	TEST WEIGHT LB/BU	YIELD[x] BU/ACRE
Nontreated control	1.5 a	1.5 a	0.0	12.6	55.6	47.1
Delaro 325 SC 8.0 fl oz at R3	0.3 c	0.4 b	0.0	12.8	55.4	48.8
Miravis Neo 2.5 SE 13.7 fl oz at R3	0.1 c	0.3 b	0.5	12.6	55.7	48.1
Lucento 4.17 SC 5.0 fl oz at R3	0.2 c	0.4 b	0.3	12.7	55.8	47.0
Priaxor Xemium 4.0 fl oz at R3	1.3 ab	0.8 ab	0.0	12.9	56.0	50.2
Trivapro 2.21 SE 13.7 fl oz at R3	0.5 c	0.3 b	0.3	12.7	55.8	48.8
Delaro 325 SC 8.0 fl oz at R5	0.2 c	0.3 b	0.0	12.6	55.7	49.8
Miravis Neo 2.5 SE 13.7 fl oz at R5	0.6 bc	0.0 b	0.0	12.5	55.8	50.6
Lucento 4.17 SC 5.0 fl oz at R5	0.0 c	0.1 b	0.0	12.6	55.6	49.5
Priaxor Xemium 4.0 fl oz at R5	0.4 c	0.4 b	0.3	13.3	56.0	48.8
Trivapro 2.21 SE 13.7 fl oz at R5	0.4 c	0.8 ab	0.3	12.7	55.9	46.5
P-value[w]	*0.0030*	*0.1066*	*0.6236*	*0.0160*	*0.1618*	*0.5718*

[z] Fungicide treatments were applied on August 5 at the beginning pod (R3) growth stage, and all treatments contained a nonionic surfactant (Preference) at a rate of 0.25% v/v.

[y] Foliar disease severity was rated on a scale of 0–100% in the canopy on September 20. FLS = frogeye leaf spot, SBS = Septoria brown spot, CLB = Cercospora leaf blight.

[x] Yields were adjusted to 13.0% moisture at harvest on October 17.

[w] Means followed by the same letter are not significantly different based on Fisher's Least Significant Difference test (LSD; α=0.05).

FUSARIUM HEAD BLIGHT (FHB) INTEGRATED FUNGICIDE TRIALS ON WHEAT IN CENTRAL INDIANA, 2019 (WHT19-01.ACRE)

S. Shim, J. D. Ravellette, and D. E. P. Telenko, Department of Botany and Plant Pathology, Purdue University West Lafayette, Indiana 47907-2054

WHEAT (*TRITICUM AESTIVUM* P25R40)

Fusarium head blight, *Fusarium graminearum*
Leaf blotch, *Stagnospora nodorum*
Leaf rust, *Puccinia triticina*

A trial was established at the Purdue Agronomy Center for Research and Education (ACRE) in Tippecanoe County, IN. The experiment was a randomized complete block design with four replications. Plots were 7.5 feet wide and 20 feet long and consisted of 12 rows spaced 7.5 inches apart, and the center of each plot was used for evaluation. The previous crop was corn. Prior to planting, the field was vertically tilled twice on September 20, 2018, and disked and cultivated on October 1, 2018. Nitrogen 28% was applied at 30 lb/acre on March 27, 2019. On October 3, 2018, wheat cultivar P25R40 was drilled at 7.5-inch spacing. Harmony Extra at 0.8 oz/acre plus AMS at 1 lb/acre plus NIS at 0.25% v/v was applied on April 27, 2019, for weed management. Fungicide applications were applied with a CO_2 pressurized backpack sprayer using a handheld boom fitted with a pair of TJ8001VS nozzles spaced 20 inches apart and directed forward and backward at a 45 degree angle, which delivered 10 gal/acre at 40 psi. Fungicides were applied on May 23 at the Feekes 10.3, May 29 at the Feekes 10.5.1, and June 3 at the Feekes 10.5.4 growth stages. All plots were inoculated with a mixture of isolates of *Fusarium graminearum* endemic to Indiana on May 31 at a concentration of 50,000 spores/ml. Spore suspension was applied at 300 ml/plot with the CO_2 handheld sprayer described previously. Disease ratings were assessed on June 25. Fusarium head blight (FHB) disease incidence (DI) was measured as the number of infected heads out of 100 in each plot and calculated as a percentage. FHB disease severity (DS) was rated by visually assessing the percentage of those infected heads. FHB index was calculated as (FHB % DI/Average FHB % DS)/100 per plot. Disease severity of leaf blotch was rated by visually assessing the percentage of symptomatic tissue on five flag leaves per plot for leaf blotch and five heads per plot for glume blotch. Values for each plot were averaged before analysis. The eight center rows of each plot were harvested with a Kincaid plot combine on July 9, and yields were adjusted to 13.5% moisture. Data were subjected to analysis of variance (SAS 9.4, 2019), and means were compared using Fisher's Least Significant Difference test (LSD; α=0.05).

In 2019, weather conditions were favorable for FHB, leaf blotch and leaf rust diseases. FHB was the most prominent disease in the trial. All fungicides reduced both % FHB DI and % FHB DS and severity of leaf blotch and leaf rust over the nontreated control on June 25 except Caramba for % FHJB DI (Table 12). Miravis Ace applied at Feekes 10.5.1 followed by Caramba at 10.5.4 resulted in the lowest FHB Index on June 25, but this treatment was not different from Miravis Ace applied alone at 10.3 or 10.5.1 and the Miravis Ace at 10.5.1 followed by Prosaro at 10.5.4. The concentration of the mycotoxin deoxynivalenol (DON) was significantly reduced over the nontreated control with Caramba at 10.5.1, Miravis Ace at 10.5.1, and Miravis Ace at 10.5.1

followed by Prosaro at 10.5.4 (Table 13). The percentage of Fusarium damaged kernels (FDK) was significantly reduced with all treatments over nontreated the control except Prosaro at 10.5.1 and Miravis Ace at 10.5.4. All fungicide treatments significantly increased yield over the nontreated control except for Miravis Ace at 10.3. Miravis Ace at 10.5.1 followed by Caramba at 10.5.4 resulted in the greatest yield.

TABLE 12. ***Effect of Fungicide on Fusarium Head Blight (FHB) and Foliar Disease in Wheat***

TREATMENT, RATE/ACRE, AND APPLICATION TIMING[Z]	FHB % DI[Y]	FHB % DS[Y]	FHB INDEX[X]	LEAF BLOTCH[W] %	LEAF RUST[W] %
Nontreated control	68.8 a	60.6 a	42.2 a	33.8 a	1.8 a
Prosaro 6.5 fl oz at 10.5.1	50.5 bc	31.7 b	16.1 b	12.7 b	0.0 b
Caramba 13.5 fl oz at 10.5.1	60.3 ab	28.9 bc	17.5 b	7.5 b	0.1 b
Miravis Ace 5.2SC 13.7 fl oz at 10.3	46.3 bcd	28.2 bc	13.2 bc	7.3 b	0.3 b
Miravis Ace 5.2SC 13.7 fl oz at 10.5.1	45.3 cd	24.0 bc	10.9 bc	8.4 b	0.0 b
Miravis Ace 5.2SC 13.7 fl oz at 10.5.4	47.0 bcd	31.0 b	15.1 b	5.1 b	0.2 b
Miravis Ace 5.2SC 13.7 fl oz at 10.5.1 fb Prosaro 6.5 fl oz at 10.5.4	42.5 cd	20.2 cd	9.2 bc	5.9 b	0.2 b
Miravis Ace 5.2SC 13.7 fl oz at 10.5.1 fb Caramba 13.5 fl oz at 10.5.4	35.0 d	12.6 d	4.4 c	5.2 b	0.1 b
P-value[V]	*0.0026*	*0.0001*	*0.0001*	*0.0001*	*0.0001*

[Z] All treatments contained a nonionic surfactant (Preference) at a rate of 0.125% v/v. Plots were inoculated with *Fusarium graminearum* spore suspension (40,000–100,000 spores/ml) 24–26 hours after the treatment at Feekes 10.5.1. Spore suspension was applied at 300 ml/plot with a handheld sprayer on May 31. fb = followed by.

[Y] Fusarium head blight (FHB) disease incidence (DI) was measured as the number of infected heads out of 100 in each plot and calculated as a percentage. FHB disease severity (DS) was rated by visually assessing the percentage of those infected heads on June 25.

[X] FHB index was calculated as (FHB % DI/Average FHB % DS)/100 per plot.

[W] Disease severity of leaf blotch and leaf rust was rated by visually assessing the percentage of symptomatic tissue on five flag leaves per plot for leaf blotch and five heads per plot for glume blotch on June 25.

[V] Means followed by the same letter are not significantly different based on Fisher's Least Significant Difference test (LSD; α=0.05).

TABLE 13. *Effect of Fungicide on Deoxynivalenol (DON), Fusarium Damaged Kernels (FDK), and Wheat Yield*

TREATMENT, RATE/ACRE, AND APPLICATION TIMING[z]	DON[y] PPM	FDK[x] %	HARVEST MOISTURE %	TEST WEIGHT LB/BU	YIELD[w] BU/ACRE
Nontreated control	5.2 ab	23.8 a	12.0 f	56.4 d	69.8 c
Prosaro 6.5.0 fl oz at 10.5.1	4.5 abc	18.8 ab	12.3 e	57.7 cd	80.1 ab
Caramba 13.5.0 fl oz at 10.5.1	3.3 bc	9.3 cd	12.4 de	58.5 abc	78.2 b
Miravis Ace 5.2SC 13.7 fl oz at 10.3	5.8 a	14.5 bc	12.5 bcd	57.9 bc	77.2 bc
Miravis Ace 5.2SC 13.7 fl oz at 10.5.1	3.4 bc	8.5 cd	12.6 bc	59.7 a	83.7 ab
Miravis Ace 5.2SC 13.7 fl oz at 10.5.4	4.8 ab	17.5 ab	12.5 cd	58.1 bc	80.5 ab
Miravis Ace 5.2SC 13.7 fl oz at 10.5.1 fb Prosaro 6.5.0 fl oz at 10.5.4	2.6 c	6.5 d	12.8 a	59.3 ab	84.5 ab
Miravis Ace 5.2SC 13.7 fl oz at 10.5.1 fb Caramba 13.5.0 fl oz at 10.5.4	3.2 bc	7.8 cd	12.7 ab	59.5 a	87.2 a
P-value[v]	0.0009	0.0392	0.0001	0.0009	0.0090

[z] All treatments contained a nonionic surfactant (Preference) at a rate of 0.125% v/v. Plots were inoculated with *Fusarium graminearum* spore suspension (40,000–100,000 spores/ml) 24–26 hours after the treatment at Feekes 10.5.1. Spore suspension were applied at 300 ml/plot with a handheld sprayer on May 31. fb = followed by.

[y] Analysis of the mycotoxin deoxynivalenol (DON) was completed by the University of Minnesota DON Testing Lab.

[x] FDK = percentage of Fusarium damaged kernels of subsamples taken from each plot.

[w] Yields were adjusted to 13.5% moisture at harvest on July 9.

[v] Means followed by the same letter are not significantly different based on Fisher's Least Significant Difference test (LSD; α=0.05).

EVALUATION OF FOLIAR FUNGICIDES FOR WHEAT DISEASE MANAGEMENT IN CENTRAL INDIANA, 2019 (WHT-19-03.ACRE)

S. Shim, J. D. Ravellette, and D. E. P. Telenko, Department of Botany and Plant Pathology, Purdue University
West Lafayette, Indiana 47907-2054

WHEAT (*TRITICUM AESTIVUM* P25R40)

Fusarium head blight, *Fusarium graminearum*
Stagnospora leaf and glume blotch, *Stagnospora nodorum*

A trial was established at the Purdue Agronomy Center for Research and Education (ACRE) in Tippecanoe County, IN. The experiment was a randomized complete block design with four replications. Plots were 7.5 feet wide and 20 feet long and consisted of 12 rows spaced 7.5 inches apart, and the center of each plot was used for evaluation. The previous crop was corn. Prior to planting, the field was vertically tilled twice on September 20, 2018, and disked and cultivated on October 1, 2018. Nitrogen 28% was applied at 30 lb/acre on March 27, 2019. On October 3, 2018, wheat cultivar P25R40 was drilled at 7.5-inch spacing. Harmony Extra at 0.8 oz/acre plus AMS at 1 lb/acre plus NIS at 0.25% v/v was applied on April 27, 2019, for weed management. Fungicide applications were applied with a CO_2 pressurized backpack sprayer using a handheld boom fitted with a pair of TJ8001VS nozzles spaced 20 inches apart and directed forward and backward at a 45 degree angle, which delivered 10 gal/acre at 40 psi. Fungicides were applied on May 29 at the Feekes growth stage 10.5.1. All plots were inoculated with a mixture of isolates of *Fusarium graminearum* endemic to Indiana on May 31. The spore suspension (50,000 spores/ml) was applied at 300 ml/plot with the CO_2 handheld sprayer described previously. Disease ratings were assessed on June 18. Fusarium head blight (FHB) disease incidence (DI) was measured as the number of infected heads out of 100 in each plot and calculated as a percentage. FHB disease severity (DS) was rated by visually assessing the percentage of those infected heads. FHB index was calculated as (FHB % DI/Average FHB % DS)/100 per plot. Disease severity of Stagnospora leaf and glume blotch was rated by visually assessing the percentage of symptomatic leaf tissue on five flag leaves per plot for leaf blotch and five heads per plot for glume blotch. Values for each plot were averaged before analysis. The eight center rows of each plot were harvested with a Kincaid plot combine on July 9, and yields were adjusted to 13.5% moisture. Data were subjected to analysis of variance (SAS 9.4, 2019), and means were compared using Fisher's Least Significant Difference test (LSD; α=0.05).

In 2019, weather conditions were favorable for FHB, leaf blotch, and glume blotch diseases. FHB was the most prominent disease. The percentage of FHB DI was reduced by Prosaro, Caramba, Miravis Ace, and USF0115 over the nontreated control on June 18 (Table 14). The percentages of FHB DS and FHB Index were reduced by all fungicide treatments over the nontreated control except Stratego YLD. FHB Index was lowest with Miravis Ace, but this was not significantly different from Prosaro, Caramba, and USF0115. All fungicide treatments significantly reduced the percentage of leaf blotch and glume blotch over the nontreated control except for Stratego YLD. The concentration of deoxynivalenol (DON) was significantly reduced over the nontreated control for all treatments except for Trivapro and Tilt (Table 15). All fungicide treatments significantly increased yield over the nontreated control except for Headline. Miravis Ace resulted in the highest yield but was not significantly different from USF0115.

TABLE 14. ***Effect of Fungicide on Fusarium Head Blight (FHB) and Foliar Disease in Wheat***

TREATMENT AND RATE/ACRE[z]	FHB % DI[y]	FHB % DS[y]	FHB INDEX[x]	LEAF BLOTCH[v] %	GLUME BLOTCH[v] %
Nontreated control	85.3 a	37.6 a	32.3 a	32.4 a	27.5 a
Prosaro 8.2 fl oz	71.3 bc	22.2 bcd	16.2 cde	16.2 cde	7.8 bc
Caramba 13.5 fl oz	72.0 bc	19.0 cd	13.9 de	14.0 de	4.5 c
Miravis Ace 5.2 SC 13.7 fl oz	67.8 c	15.5 d	10.3 e	10.3 e	5.0 c
USF0115 10.3 fl oz	71.0 bc	21.0 bcd	14.8 cde	14.9 cde	7.7 bc
Trivapro 2.21 SE 13.7 fl oz	80.0 ab	27.0 b	21.5 bc	21.5 bc	6.0 c
Stratego YLD 4.18 SC 4.0 fl oz	78.8 ab	35.7 a	27.5 ab	27.5 ab	9.5 bc
Delaro 325 SC 8.0 fl oz	76.3 abc	25.5 bc	19.4 cd	19.4 cd	8.2 bc
Tilt 3.6 EC 4.0 fl oz	85.3 a	24.5 bc	20.9 bc	20.9 bc	15.4 b
Headline 2.09 SC 9.0 fl oz	80.3 ab	25.1 bc	20.1 cd	20.2 cd	4.4 c
P-value[u]	*0.0001*	*0.0001*	*0.0001*	*0.0001*	*0.0001*

[z] Fungicide treatments were applied at Feekes 10.5.1. All treatments contained a nonionic surfactant (Preference) at a rate of 0.125% v/v. Plots were inoculated with *Fusarium graminearum* spore suspension (40,000–100,000 spores/ml) 24–26 hours after the treatment at Feekes 10.5.1. Spore suspension was applied at 300 ml/plot with a handheld sprayer on May 31.

[y] Fusarium head blight (FHB) disease incidence (DI) was measured as the number of infected heads out of 100 in each plot and calculated as a percentage on June 18. FHB disease severity (DS) was rated by visually assessing the percentage of those infected heads.

[x] FHB index was calculated as (FHB % DI/Average FHB % DS)/100 per plot.

[v] Disease severity of Stagnospora leaf and glume blotch was rated by visually assessing the percentage of symptomatic leaf tissue on five flag leaves per plot for leaf blotch and five heads per plot for glume blotch on June 18.

[u] Means followed by the same letter are not significantly different based on Fisher's Least Significant Difference test (LSD; α=0.05).

TABLE 15. ***Effect of Fungicide on Fusarium Damaged Kernels (FDK), Deoxynivalenol (DON), and Wheat Yield***

TREATMENT AND RATE/ACRE[z]	FDK[Y] %	DON[X] PPM	HARVEST MOISTURE %	TEST WEIGHT LBS/BU	YIELD[W] BU/ACRE
Nontreated control	23.8	7.98 a	11.8 b	55.4 d	66.3 d
Prosaro 8.2 fl oz	13.8	3.68 d	12.3 a	57.5 bc	77.6 bc
Caramba 13.5 fl oz	11.5	3.50 d	12.3 a	57.3 bc	76.4 bc
Miravis Ace 5.2 SC 13.7 fl oz	11.0	3.05 d	12.3 a	59.0 a	85.2 a
USF0115 10.3 fl oz	12.8	3.48 d	12.1 ab	58.1 ab	82.1 ab
Trivapro 2.21 SE 13.7 fl oz	20.8	6.65 abc	12.1 ab	56.5 c	74.1 c
Stratego YLD 4.18 SC 4.0 fl oz	18.8	5.28 c	12.1 a	56.6 bc	77.4 bc
Delaro 325 SC 8.0 fl oz	17.5	3.47 d	12.2 a	57.1 c	74.3 c
Tilt 3.6 EC 4.0 fl oz	23.8	6.90 ab	12.1 ab	56.7 c	74.9 c
Headline 2.09 SC 9.0 fl oz	17.5	5.88 bc	12.3 a	56.4 de	72.1 ed
P-value[v]	0.0530	*0.0001*	*0.0343*	*0.0001*	*0.0001*

[z] Fungicide treatments were applied at Feekes 10.5.1. All treatments contained a nonionic surfactant (Preference) at a rate of 0.125% v/v. Plots were inoculated with *Fusarium graminearum* spore suspension (40,000–100,000 spores/ml) 24–26 hours after the treatment at Feekes 10.5.1. Spore suspension was applied at 300 ml/plot with a handheld sprayer on May 31.

[y] FDK = percentage of Fusarium damaged kernels of subsamples taken from each plot.

[x] Analysisof the mycotoxin deoxynivalenol (DON) wa completed by the University of Minnesota DON Testing Lab.

[w] Yields were adjusted to 13.5% moisture at harvest on July 9.

[v] Means followed by the same letter are not significantly different based on Fisher's Least Significant Difference test (LSD; α=0.05).

PINNEY PURDUE AGRICULTURAL CENTER (PPAC)

UNIFORM FUNGICIDE COMPARISON FOR TAR SPOT OF CORN IN NORTHWESTERN INDIANA, 2019 (COR19-03.PPAC)

A. Chaille, S. Shim, J. D. Ravellette, and D. E. P. Telenko, Department of Botany and Plant Pathology, Purdue University West Lafayette, Indiana 47907-2054

CORN (*ZEA MAYS* W2585SSRIB)

Tar spot, *Phyllachora maydis*

A trial was established at the Pinney Purdue Agricultural Center (PPAC) in Porter County, IN. The experiment was a randomized complete block design with four replications. Plots were 10 feet wide and 30 feet long and consisted of four rows, and the two center rows were used for evaluation. The previous crop was corn. Standard practices for grain corn production in Indiana were followed. Corn hybrid W2585SSRIB was planted in 30-inch row spacing at a rate of 34,000 seeds/acre on June 8. The field was overhead irrigated weekly at 1 inch unless weekly rainfall was 1 inch or higher to encourage disease. All fungicide applications were applied at 15 gal/acre and 40 psi using a Lee self-propelled sprayer equipped with a 10-foot boom, fitted with six TJ-VS 8002 nozzles spaced 20 inches apart, at 3.6 mph. Fungicides were applied on August 8 at the tassel/silk (VT/R1) growth stage. Disease ratings were assessed on September 21 and on September 30 at the milk (R3) and dent (R5) growth stages, respectively. Tar spot was rated by visually assessing the percentage of stroma and the percentage of symptomatic tissues (chlorosis and necrosis) per leaf on five plants in each plot at the ear leaf (EL), ear leaf minus two (EL−2) leaves, and ear leaf plus two (EL+2) leaves. The values of the five leaves for each plot were averaged before analysis. The two center rows of each plot were harvested on October 28, and yields were adjusted to 15.5% moisture. Data were subjected to analysis of variance (SAS 9.4, 2019), and means were compared using Tukey-Kramer (α=0.05).

In 2019, weather conditions were favorable for disease. Tar spot was the most prominent diseases in the trial and reached moderate to high severity. All fungicides significantly reduced the percentage of stroma on the (EL−2) leaves, the EL leaf, and the (EL+2) leaves and the percentage of chlorotic and necrotic symptoms of tar spot on the (EL−2)leaves and the EL over the nontreated control on September 21 (Table 16). All fungicides significantly reduced the percentage of stroma and chlorotic and necrotic symptoms on the (EL−2),

EL, and (EL+2) compared to the nontreated control on September 30 (Table 17). Headline had the lowest percentage of stroma on the ear leaf minus two on September 30 but was not significantly different from Veltyma and Delaro. All fungicide treatments significantly increased the percentage of canopy greenness over the nontreated control on September 30 (Table 18). No significant treatment effects were detected for lodging, test weight, and corn yield.

TABLE 16. ***Effect of Fungicide on Tar Spot on September 21***

TREATMENT AND RATE/ACRE[z]	TAR SPOT % STROMA[y] EL–2	TAR SPOT % STROMA[y] EL	TAR SPOT % STROMA[y] EL+2	TAR SPOT % CHLOR/NEC[x] EL–2	TAR SPOT % CHLOR/NEC[x] EL	TAR SPOT % CHLOR/NEC[x] EL+2
Nontreated control	7.2 a	2.8 a	1.7 a	7.6 a	0.8 a	0.1
Trivapro 2.21 SE 13.7 fl oz	3.0 b	1.3 b	1.0 cd	2.2 b	0.1 b	0.0
Miravis Neo 2.5 SE 13.7 fl oz	3.2 b	1.3 b	1.0 d	1.0 b	0.1 b	0.1
Delaro 325 SC 12.0 fl oz	1.5 b	1.3 b	1.2 bcd	0.3 b	0.0 b	0.0
Headline AMP 1.68 SC 14.4 fl oz	1.5 b	1.2 b	1.1 cd	0.3 b	0.0 b	0.1
Topguard EQ 7.0 fl oz	2.7 b	1.3 b	1.4 b	1.3 b	0.0 b	0.1
Quilt Xcel 2.2 SE 14 fl oz	2.9 b	1.3 b	1.1 cd	2.5 b	0.0 b	0.0
Veltyma 3.34 SC 7.0 fl oz	2.0 b	1.3 b	1.2 bc	0.2 b	0.0 b	0.1
Aproach Prima 2.34 SC 6.8 fl oz	2.6 b	1.1 b	1.0 cd	1.5 b	0.0 b	0.0
Headline 2.09 SC 12.0 fl oz	1.4 b	1.1 b	1.1 cd	0.1 b	0.0 b	0.3
Proline 480 SC 5.7 fl oz	3.2 b	1.4 b	1.1 cd	1.6 b	0.1 b	0.5
P-value[w]	*0.0001*	*0.0001*	*0.0001*	*0.0002*	*0.0001*	*0.4446*

[z] Fungicide treatments were applied on August 8 at the tassel/silk (VT/R1) growth stage, and all treatments contained a nonionic surfactant (Preference) at a rate of 0.25% v/v.

[y] Tar spot stroma was visually assessed as a percentage (0–100%) of leaf area on five plants in each plot at the ear leaf (EL), ear leaf minus two (EL–2), and ear leaf plus two (EL+2).

[x] Tar spot chlorosis and necrosis symptoms were visually assessed as a percentage (0–100%) of leaf area on five plants in each plot at the EL, EL–2, and EL+2.

[w] Means followed by the same letter are not significantly different based on Tukey's-Kramer (α=0.05).

TABLE 17. ***Effect of Fungicide on Tar Spot on September 30***

TREATMENT AND RATE/ACRE[z]	TAR SPOT % STROMA[y] EL–2	TAR SPOT % STROMA[y] EL	TAR SPOT % STROMA[y] EL+2	TAR SPOT % CHLOR/NEC[x] EL–2	TAR SPOT % CHLOR/NEC[x] EL	TAR SPOT % CHLOR/NEC[x] EL+2
Nontreated control	38.8 a	34.3 a	23.5 a	77.7 a	57.8 a	32.4 a
Trivapro 2.21 SE 13.7 fl oz	26.5 b	16.0 b	7.9 b	57.0 b	24.0 b	5.7 b
Miravis Neo 2.5 SE 13.7 fl oz	24.8 bc	16.8 b	7.6 bc	56.3 b	18.2 b	5.1 b
Delaro 325 SC 12.0 fl oz	14.8 def	9.8 bc	6.1 cd	40.5 b	9.5 b	3.9 b
Headline AMP 1.68 SC 14.4 fl oz	16.3 de	11.5 bc	6.0 cd	34.5 b	10.5 b	2.9 b
Topguard EQ 7.0 fl oz	25.5 bc	15.5 bc	6.8 bcd	52.5 b	18.5 b	4.8 b
Quilt Xcel 2.2 SE 14 fl oz	22.3 bcd	16.3 b	9.3 bc	55.1 b	17.5 b	9.2 b
Veltyma 3.34 SC 7.0 fl oz	13.4 ef	12.0 bc	6.1 cd	35.7 b	14.8 b	5.4 b
Aproach Prima 2.34 SC 6.8 fl oz	18.3 cde	11.8 bc	5.7 cd	52.5 b	11.2 b	3.2 b
Headline 2.09 SC 12.0 fl oz	8.1 f	7.0 c	4.3 d	21.3 b	9.2 b	2.0 b
Proline 480 SC 5.7 fl oz	25.5 bc	16.3 b	10.2 b	65.5 b	24.3 b	9.8 b
P-value[w]	*0.0001*	*0.0001*	*0.0001*	*0.0001*	*0.0001*	*0.0001*

[z] Fungicide treatments were applied on August 8 at the tassel/silk (VT/R1) growth stage, and all treatments contained a nonionic surfactant (Preference) at a rate of 0.25% v/v.

[y] Tar spot stroma was visually assessed as a percentage (0–100%) of leaf area on five plants in each plot at the ear leaf (EL), ear leaf minus two (EL–2), ear leaf plus two (EL+2).

[x] Tar spot chlorosis and necrosis symptoms were visually assessed as a percentage (0–100%) of leaf area on five plants in each plot at the EL, EL–2, and EL+2.

[w] Means followed by the same letter are not significantly different based on Tukey's-Kramer (α=0.05).

TABLE 18. ***Effect of Fungicide on Canopy Greenness, Lodging, and Corn Yield***

TREATMENT AND RATE/ACRE[z]	CANOPY GREEN[y] %	LODGING[x] %	HARVEST MOISTURE %	TEST WEIGHT LB/BU	YIELD[w] BU/ACRE
Nontreated control	33.4 d	6.6	22.8 d	51.2	202.7
Trivapro 2.21 SE 13.7 fl oz	50.0 c	0.0	23.6 bcd	50.2	212.2
Miravis Neo 2.5 SE 13.7 fl oz	53.8 bc	0.0	23.9 a-d	49.9	215.3
Delaro 325 SC 12.0 fl oz	62.5 ab	0.0	24.4 abc	50.4	223.6
Headline AMP 1.68 SC 14.4 fl oz	65.0 a	0.0	24.6 ab	50.3	218.1
Topguard EQ 7.0 fl oz	58.8 abc	0.0	24.1 abc	50.0	211.8
Quilt Xcel 2.2 SE 14 fl oz	57.5 abc	0.0	24.1 abc	50.7	214.2
Veltyma 3.34 SC 7.0 fl oz	61.3 ab	0.0	24.8 ab	50.2	215.4
Aproach Prima 2.34 SC 6.8 fl oz	62.5 ab	0.0	24.4 abc	49.9	209.2
Headline 2.09 SC 12.0 fl oz	66.3 a	0.0	25.0 a	50.4	214.0
Proline 480 SC 5.7 fl oz	50.0 c	2.5	23.2 cd	50.6	206.1
P-value[w]	*0.0001*	*0.3399*	*0.0001*	*0.0558*	*0.1502*

[z] Fungicide treatments were applied on August 8 at the tassel/silk (VT/R1) growth stage, and all treatments contained a nonionic surfactant (Preference) at a rate of 0.25% v/v.

[y] Canopy greenness was visually assessed as a percentage (0–100%) of canopy on September 30.

[x] Lodging = percentage of lodged stalks when pushed from shoulder height to 45° from vertical on September 30.

[w] Yields were adjusted to 15.5% moisture at harvest on October 28.

[v] Means followed by the same letter are not significantly different based on Tukey's-Kramer (α=0.05).

UNIFORM FUNGICIDE TIMING AND TAR SPOT MODEL VALIDATION IN CORN IN NORTHWESTERN INDIANA, 2019 (COR19-05.PPAC)

T. Ross, S. Shim, J. D. Ravellette, and D. E. P. Telenko, Department of Botany and Plant Pathology, Purdue University West Lafayette, Indiana 47907-2054

CORN (*ZEA MAYS* W2585SSRIB)

Tar spot; *Phyllachora maydis*

A trial was established at the Pinney Purdue Agricultural Center (PPAC) in Porter County, IN. The experiment was a randomized complete block design with four replications. Plots were 10 feet wide and 30 feet long and consisted of four rows, and the two center rows were used for evaluation. The previous crop was corn. Standard practices for nonirrigated grain corn production in Indiana were followed. Corn hybrid W2585SSRIB was planted in 30-inch row spacing at a rate of 34,000 seeds/acre on June 8. All fungicide applications were applied at 15 gal/acre and 40 psi using a Lee self-propelled sprayer equipped with a 10-foot boom, fitted with six TJ-VS 8002 nozzles spaced 20 inches apart, at 3.6 mph. Fungicides were applied on July 8, July 15, July 19, August 7, and August 23 at the seven-leaf (V7), nine-leaf (V9), 10-leaf (V10), tassel/silk (VT/R1), and blister (R2) growth stages, respectively. A prediction model-based treatment was included in the trial, but the model never triggered a fungicide application during the season at PPAC; therefore, this treatment provided an additional nontreated control for comparison. Disease ratings were assessed on September 18 at the early dent (R5) growth stage and on September 30 at the late R5 growth stage. Tar spot was rated by visually assessing the percentage of stroma and the percentage of symptomatic tissues (chlorosis and necrosis) per leaf on five plants in each plot at the ear leaf, ear leaf minus two, and ear leaf plus two. Values for each plot were averaged before analysis. The two center rows of each plot were harvested on October 28, and yields were adjusted to 15.5% moisture. Data were subjected to analysis of variance (SAS 9.4, 2019), and means were compared using Fisher's Least Significant Difference test (LSD; α=0.05).

In 2019, weather conditions were favorable for disease. Tar spot was first detected in the trial on August 2 and was the most prominent disease in the trial, reaching moderate to high severity. Trivapro applied at the V9, V10, VT/R1, R2, and V7 growth stages followed by VT significantly reduced tar spot stroma over the nontreated control on the ear leaf minus two and the ear leaf on September 18, but no differences were detected on ear leaf plus two (Table 19). Chlorotic and necrotic symptoms of tar spot on the ear leaf minus two and the ear leaf on September 18 were reduced by Trivapro applied at the VT/R1, R2, and V7 growth stages followed by VT application timings. All application timings reduced chlorotic and necrotic symptoms over the nontreated control on the ear leaf plus two on September 18 except the VT/R1 timing. By September 30, tar spot stroma was only significantly less in treatments made at VT/R1 or R2 compared to the nontreated control for all leaf ratings (Table 20). The VT/R1 and R2 applications also significantly reduced chlorotic and necrotic symptoms on all leaves on September 30. Trivapro applied at V10, VT, R2, and V7 followed by VT significantly increased the percent of canopy greenness of the corn over the nontreated control on both September 18 and September 30 (Table 21). No differences in lodging were detected between treatments. Trivapro applied at VT/R1, R2, and V7 followed by VT increased corn yield over the nontreated control.

TABLE 19. ***Effect of Fungicide on Tar Spot on September 18***

TREATMENT, RATE/ACRE, AND TIMING[Z]	TAR SPOT % STROMA[Y] EL−2	TAR SPOT % STROMA[Y] EL	TAR SPOT % STROMA[Y] EL+2	TAR SPOT % CHLOR/ NEC[X] EL−2	TAR SPOT % CHLOR/ NEC[X] EL	TAR SPOT % CHLOR/ NEC[X] EL+2
Nontreated control	28.0 a	6.8 a	2.6	17.4 a	1.6 ab	0.2 a
Trivapro 2.21 SE 13.7 fl oz at V7	25.8 ab	4.8 abc	2.0	19.8 a	1.1 abc	0.0 c
Trivapro 2.21 SE 13.7 fl oz at V9	22.6 b	4.3 bc	1.7	11.1 ab	0.9 bcd	0.0 c
Trivapro 2.21 SE 13.7 fl oz at V10	8.5 c	3.8 cd	1.7	2.5 bc	0.2 cd	0.2 ab
Trivapro 2.21 SE 13.7 fl oz at VT/R1	4.7 cd	2.9 cde	1.0	1.3 c	0.0 d	0.0 c
Trivapro 2.21 SE 13.7 fl oz at R2	4.2 d	0.7 e	0.6	2.2 bc	0.1 cd	0.1 bc
Trivapro 2.21 SE 13.7 fl oz V7 fb VT	5.8 cd	1.7 de	2.0	1.6 c	0.3 cd	0.0 c
Trivapro 2.21 SE 13.7 fl oz at tar spot model	25.6 ab	6.2 ab	2.4	17.8 a	2.0 a	0.1 abc
P-value[W]	*0.0001*	*0.0002*	*0.0821*	*0.0002*	*0.0036*	*0.0389*

[z] Fungicide treatments were applied on July 8, July 15, July 19, August 7, and August 23 at the V7 (seven-leaf), V9 (nine-leaf), V10 (10-leaf), VT/R1 (silk), and R2 (blister) growth stages, respectively. All treatments contained a nonionic surfactant (Preference) at a rate of 0.25% v/v. fb = followed by. Model based application was not applied.

[y] Tar spot stroma was visually assessed as a percentage (0–100%) of leaf area on five plants in each plot at the ear leaf (EL), ear leaf minus two (EL−2), ear leaf plus two (EL+2).

[x] Tar spot chlorosis and necrosis symptoms were visually assessed as a percentage (0–100%) of leaf area on five plants in each plot at the EL, EL−2, and EL+2.

[w] Means followed by the same letter are not significantly different based on Fisher's Least Significant Difference test (LSD; α=0.05).

TABLE 20. ***Effect of Fungicide on Tar Spot on September 30***

TREATMENT, RATE/ACRE, AND TIMING[z]	TAR SPOT % STROMA[y] EL–2	TAR SPOT % STROMA[y] EL	TAR SPOT % STROMA[y] EL+2	TAR SPOT % CHLOR/ NEC[x] EL–2	TAR SPOT % CHLOR/ NEC[x] EL	TAR SPOT % CHLOR/ NEC[x] EL+2
Nontreated control	37.8 ab	38.8 a	30.3 ab	88.5 ab	75.0 ab	45.0 ab
Trivapro 2.21 SE 13.7 fl oz at V7	41.8 a	39.0 a	36.5 a	95.8 a	80.5 a	53.5 a
Trivapro 2.21 SE 13.7 fl oz at V9	36.3 ab	35.0 a	28.8 ab	78.5 bc	59.3 bc	28.4 bc
Trivapro 2.21 SE 13.7 fl oz at V10	31.8 bc	34.8 a	26.8 bc	70.0 cd	56.3 bc	28.5 bc
Trivapro 2.21 SE 13.7 fl oz at VT/R1	24.3 d	23.8 b	16.5 d	65.3 d	43.0 cd	16.8 cd
Trivapro 2.21 SE 13.7 fl oz at R2	23.3 d	23.5 b	12.1 d	75.8 cd	30.0 d	7.0 d
Trivapro 2.21 SE 13.7 fl oz V7 fb VT	25.3 cd	24.5 b	19.8 cd	68.8 cd	42.3 cd	16.3 cd
Trivapro 2.21 SE 13.7 fl oz at tar spot model	39.3 a	38.0 a	31.0 ab	94.0 a	72.5 a	53.8 a
P-value[w]	*0.0001*	*0.0001*	*0.0001*	*0.0001*	*0.0001*	*0.0001*

[z] Fungicide treatments were applied on July 8, July 15, July 19, August 7, and August 23 at the V7 (seven-leaf), V9 (nine-leaf), V10 (10-leaf), VT/R1 (silk), and R2 (blister) growth stages, respectively. All treatments contained a nonionic surfactant (Preference) at a rate of 0.25% v/v. fb = followed by. Model based application was not applied.

[y] Tar spot stroma was visually assessed as a percentage (0–100%) of leaf area on five plants in each plot at the ear leaf (EL), ear leaf minus two (EL–2), ear leaf plus two (EL+2).

[x] Tar spot chlorosis and necrosis symptoms were visually assessed as a percentage (0–100%) of leaf area on five plants in each plot at the EL, EL–2, and EL+2.

[w] Means followed by the same letter are not significantly different based on Fisher's Least Significant Difference test (LSD; α=0.05).

TABLE 21. ***Effect of Fungicide on Canopy Greenness, Lodging, and Yield***

TREATMENT, RATE/ACRE, AND TIMINIG[z]	CANOPY GREEN[y] % SEPTEMBER 18	CANOPY GREEN[y] % SEPTEMBER 30	LODGING[x] %	HARVEST MOISTURE %	TEST WEIGHT LB/BU	YIELD[w] BU/ACRE
Nontreated control	63.8 cd	38.8 d	1.0	22.6 cd	53.3	181.9 c
Trivapro 2.21 SE 13.7 fl oz at V7	71.3 bc	41.3 cd	1.3	22.4 d	52.8	182.3 c
Trivapro 2.21 SE 13.7 fl oz at V9	75.0 b	43.8 cd	1.3	22.7 cd	52.8	188.3 bc
Trivapro 2.21 SE 13.7 fl oz at V10	86.3 a	50.0 bc	0.3	23.2 bc	52.6	191.6 abc
Trivapro 2.21 SE 13.7 fl oz at VT/R1	90.8 a	55.0 ab	1.0	23.5 ab	52.0	202.6 ab
Trivapro 2.21 SE 13.7 fl oz at R2	92.5 a	63.8 a	0.3	23.7 ab	53.4	205.2 a
Trivapro 2.21 SE 13.7 fl oz V7 fb VT	88.8 a	57.5 ab	0.3	23.9 a	51.9	202.5 ab
Trivapro 2.21 SE 13.7 fl oz at tar spot model	58.8 d	37.5 d	1.8	22.4 d	53.5	181.6 c
P-value[v]	*0.0001*	*0.0001*	*0.6175*	*0.0003*	*0.1056*	*0.0051*

[z] Fungicide treatments were applied on July 8, July 15, July 19, August 7, and August 23 at the V7 (seven-leaf), V9 (nine-leaf), V10 (10-leaf), VT/R1 (silk), and R2 (blister) growth stages, respectively. All treatments contained a nonionic surfactant (Preference) at a rate of 0.25% v/v. fb = followed by. Model based application was not applied.

[y] Canopy greenness was visually assessed as a percentage (0–100%) of canopy.

[x] Lodging = % lodged stalks when pushed from shoulder height to 45° from vertical.

[w] Yields were adjusted to 15.5% moisture at harvest on October 28.

[v] Means followed by the same letter are not significantly different based on Fisher's Least Significant Difference (LSD; α=0.05).

FUNGICIDE EVALUATION FOR FOLIAR DISEASE IN CORN IN NORTHWESTERN INDIANA, 2019 (COR19-13.PPAC)

C. Rocco da Silva, S. Shim, J. D. Ravellette, and D. E. P. Telenko, Department of Botany and Plant Pathology, Purdue University West Lafayette, Indiana 47907-2054

CORN (*ZEA MAYS* W2585SSRIB)

Tar spot, *Phyllachora maydis*

A trial was established at the Pinney Purdue Agricultural Center (PPAC) in Porter County, IN. The experiment was a randomized complete block design with four replications. Plots were 10 feet wide and 30 feet long and consisted of four rows, and the two center rows were used for evaluation. The previous crop was corn. Standard practices for corn production in Indiana were followed. Corn hybrid W2585SSRIB was planted in 30-inch row spacing at a rate of 34,000 seeds/acre on June 8. The field was overhead irrigated weekly at 1 inch unless weekly rainfall was 1 inch or higher to encourage disease. All fungicide applications were applied at 15 gal/acre and 40 psi using a Lee self-propelled sprayer equipped with a 10-foot boom, fitted with six TJ-VS 8002 nozzles spaced 20 inches apart, at 3.6 mph. Fungicides were applied on August 9 at the silk (R1) growth stage. Disease ratings were assessed on September 21 and October 2 at the dent (R5) and maturity (R6) growth stages, respectively. Tar spot was rated by visually assessing the percentage of stroma and the percentage of symptomatic tissues (chlorosis and necrosis) per leaf on five plants in each plot at the ear leaf, ear leaf minus two, ear leaf plus two. Values for each plot were averaged before analysis. The two center rows of each plot were harvested on October 25, and yields were adjusted to 15.5% moisture. Data were subjected to analysis of variance (SAS 9.4, 2019), and means were compared using Fisher's Least Significant Difference test (LSD; α=0.05).

In 2019, weather conditions were favorable for disease. Tar spot was the most prominent diseases in the trial and reached moderate to high severity. All fungicide treatments reduced severity of tar spot stroma over the nontreated control at all rating dates (Tables 22 and 23). The chlorotic and necrotic symptoms were also significantly reduced with all fungicide treatments on October 2 on the ear leaf, ear leaf minus two, and ear leaf plus two (Table 23). All fungicide treatments significantly increased the percentage of canopy greenness of the corn over the nontreated control on both August 29 and October 2 (Table 24). All fungicides significantly reduced lodging and increased and yield over the nontreated control.

TABLE 22. *Effect of Fungicide Treatment on Tar Spot on September 21*

TREATMENT AND RATE/ACRE[z]	TAR SPOT % STROMA[y] EL−2	TAR SPOT % STROMA[y] EL	TAR SPOT % STROMA[y] EL+2
Nontreated control	9.0 a	4.2 a	2.0 a
Topguard EQ 5.0 fl oz	1.8 b	1.4 b	1.0 c
Lucento 4.17 SC 5.0 fl oz	2.4 b	1.2 b	1.1 bc
VJR90-R002 7.0 fl oz	2.0 b	1.7 b	1.1 bc
VJR90-R002 8.0 fl oz	1.9 b	1.4 b	1.0 c
VJR90-R002 9.0 fl oz	2.5 b	1.4 b	1.3 bc
Miravis Neo 2.5 SE 13.7 fl oz	2.8 b	1.3 b	1.0 c
Trivapro 2.21 SE 13.7 fl oz	3.4 b	1.2 b	1.0 c
Delaro 325 SC 8.0 fl oz	1.2 b	1.1 b	1.1 bc
P-value[x]	0.0005	0.0001	0.0001

[z] Fungicide treatments were applied on August 9 at the R1 (silk) growth stage, and all treatments contained a nonionic surfactant (Preference) at a rate of 0.25% v/v.

[y] Tar spot stroma was visually assessed as a percentage (0–100%) of leaf area on five plants in each plot at the ear leaf, ear leaf minus two, and ear leaf plus two.

[x] Means followed by the same letter are not significantly different based on Fisher's Least Significant Difference test (LSD; α=0.05).

TABLE 23. *Effect of Fungicide Treatment on Tar Spot on October 2*

TREATMENT AND RATE/ACRE[z]	TAR SPOT % STROMA[Y] EL-2	TAR SPOT % STROMA[Y] EL	TAR SPOT % STROMA[Y] EL+2	TAR SPOT % CHLO/NECR[X] EL-2	TAR SPOT % CHLO/NECR[X] EL	TAR SPOT % CHLO/NECR[X] EL+2
Nontreated control	43.3 a	39.3 a	31.3 a	94.3 a	87.0 a	58.8 a
Topguard EQ 5.0 fl oz	23.3 b	17.6 b	10.4 b	65.8 b	28.5 bc	10.9 b
Lucento 4.17 SC 5.0 fl oz	31.0 b	19.3 b	10.2 b	73.3 b	36.8 b	9.4 b
VJR90-R002 7.0 fl oz	26.0 b	17.5 b	10.4 b	69.8 b	29.8 bc	15.5 b
VJR90-R002 8.0 fl oz	25.0 b	15.5 b	7.7 b	58.5 b	21.5 c	6.4 b
VJR90-R002 9.0 fl oz	28.5 b	18.6 b	10.2 b	72.5 b	36.5 b	9.4 b
Miravis Neo 2.5 SE 13.7 fl oz	25.5 b	13.6 b	7.3 b	66.5 b	27.0 bc	6.2 b
Trivapro 2.21 SE 13.7 fl oz	28.8 b	18.5 b	9.4 b	71.3 b	32.3 bc	13.5 b
Delaro 325 SC 8.0 fl oz	20.5 b	16.1 b	7.4 b	59.3 b	21.0 c	7.1 b
P-value[w]	0.0001	0.0001	0.0149	0.0320	0.0001	0.0001

[z] Fungicide treatments were applied on August 9 at the silk (R1) growth stage, and all treatments contained a nonionic surfactant (Preference) at a rate of 0.25% v/v.

[y] Tar spot stroma was visually assessed as a percentage (0–100%) of leaf area on five plants in each plot at the ear leaf (EL), ear leaf minus two (EL-2), ear leaf plus two (EL+2).

[x] Tar spot chlorosis and necrosis symptoms were visually assessed as a percentage (0–100%) of leaf area on five plants in each plot at the EL, EL-2, and EL+2.

[w] Means followed by the same letter are not significantly different based on Fisher's Least Significant Difference test (LSD; α=0.05).

TABLE 24. ***Effect of Fungicide Treatment on Canopy Greenness, Lodging, and Yield***

TREATMENT AND RATE/ACRE[z]	CANOPY GREEN[x] % SEPTEMBER 21	CANOPY GREEN[y] % OCTOBER 2	LODGING[x] %	HARVEST MOISTURE %	TEST WEIGHT LB/BU	YIELD[w] BU/ACRE
Nontreated control	77.5 b	30.0 b	17.5 a	23.1 c	51.6	197.3 b
Topguard EQ 5.0 fl oz	88.8 a	71.3 b	24.2 ab	24.5 ab	50.8	214.6 a
Lucento 4.17 SC 5.0 fl oz	90.0 a	63.8 a	2.5 b	24.1 ab	50.6	217.6 a
VJR90-R002 7.0 fl oz	90.0 a	71.3 a	0.0 b	24.4 ab	51.4	215.9 a
VJR90-R002 8.0 fl oz	92.5 a	71.3 a	2.5 b	24.3 ab	51.1	216.1 a
VJR90-R002 9.0 fl oz	90.0 a	68.8 a	2.5 b	23.7 bc	50.9	217.9 a
Miravis Neo 2.5 SE 13.7 fl oz	90.0 a	67.5 a	7.5 b	24.6 a	50.7	217.4 a
Trivapro 2.21 SE 13.7 fl oz	90.0 a	68.8 a	2.5 b	24.2 ab	50.7	217.0 a
Delaro 325 SC 8.0 fl oz	90.0 a	75.0 a	5.0 b	24.6 a	51.3	217.5 a
P-value[v]	*0.0001*	*0.0001*	*0.0111*	*0.0096*	*0.2815*	*0.0081*

[z] Fungicide treatments were applied on August 9 at the silk (R1) growth stage, and all treatments contained a nonionic surfactant (Preference) at a rate of 0.25% v/v.

[y] Canopy greenness was visually assessed as a percentage (0–100%) of canopy on September 21 and October 2.

[x] Lodging = percentage of lodged stalks when pushed from shoulder height to 45° from vertical on October 2.

[w] Yields were adjusted to 15.5% moisture at harvest on October 25.

[v] Means followed by the same letter are not significantly different based on Fisher's Least Significant Difference test (LSD; α=0.05).

FUNGICIDE COMPARISON FOR FOLIAR DISEASES IN CORN IN NORTHWESTERN INDIANA, 2019 (COR19-14.PPAC)

C. Rocca da Silva, S. Shim, J. D. Ravellette, and D. E. P. Telenko, Department of Botany and Plant Pathology, Purdue University West Lafayette, Indiana 47907-2054

CORN (*ZEA MAYS* P9998AM)

Tar spot, *Phyllachora maydis*
Gray leaf spot, *Cercospora zeae-maydis*
Common rust, *Puccinia sorghi*

A trial was established at the Pinney Purdue Agricultural Center (PPAC) in Porter County, IN. The experiment was a randomized complete block design with four replications. Plots were 10 feet wide and 30 feet long and consisted of four rows, and the two center rows were used for evaluation. The previous crop was corn. Standard practices for grain corn production in Indiana were followed. Corn hybrid P9998AM was planted in 30-inch row spacing at a rate of 34,000 seeds/acre on June 8. The field was overhead irrigated weekly at 1 inch unless weekly rainfall was 1 inch or higher to encourage disease. All fungicide applications were applied at 15 gal/acre and 40 psi using a Lee self-propelled sprayer equipped with a 10-foot boom, fitted with six TJ-VS 8002 nozzles spaced 20 inches apart, at 3.6 mph. Fungicides were applied on July 15 at the V8, July 19 at V10, and August 7 at tassel/silk (VT/R1) growth stage. Disease ratings were assessed on September 21 and October 2 at the dent (R5) and maturity (R6) growth stages, respectively. Tar spot was rated by visually assessing the percentage of stroma and the percentage of symptomatic tissues (chlorosis and necrosis) per leaf on five plants in each plot at the ear leaf (EL), ear leaf minus two (EL−2), and ear leaf plus two (EL+2). Gray leaf spot and common rust severity were rated by visually assessing the percentage of symptomatic leaf area on five plants in each plot at the ear leaf. Values for each plot were averaged before analysis. The two center rows of each plot were harvested on October 28, and yields were adjusted to 15.5% moisture. Data were subjected to analysis of variance (SAS 9.4, 2019), and means were compared using Fisher's Least Significant Difference test (LSD; α=0.05).

In 2019, weather conditions were favorable for disease. Tar spot, gray leaf spot, and common rust were the most prominent diseases in the trial and reached low severity. No differences were detected between treatments for tar spot on September 21 at all leaf levels (Table 25). All fungicides reduced gray leaf spot and common rust over the nontreated on September 21 except Dexter Xcel applied at V10. Tar spot stroma on the EL on October 2 was significantly reduced over the nontreated control by Dexter Xcel at V8 followed by VT/R1, Brixen at VT/R1, and Veltyma at VT/R1 (Table 26). No significant difference were detected between treatments for tar spot stroma severity at EL+2 or EL−2 on October 2. Tar spot chlorosis and necrosis severity on EL−2 was reduced by Dexter applied at V8 followed by VT/R1 and Dexter at V10 and Veltyma compared to the nontreated control. All fungicides reduced tar spot chlorosis and necrosis on the EL and EL+2 on October 2. All fungicide treatments significantly increased canopy greenness over the nontreated control on October 2 (Table 27). No significant differences were detected between treatment for lodging, test weight, and yield.

TABLE 25. ***Effect of Fungicide on Foliar Diseases of Corn***

TREATMENT, RATE/ACRE, AND TIMING[Z]	TAR SPOT % STROMA[Y] EL–2	TAR SPOT % STROMA[Y] EL	TAR SPOT % STROMA[Y] EL+2	GLS[X] %	CR[X] %
Nontreated control	0.9	1.2	1.0	5.8 a	5.1 a
Dexter Xcel 48.0 fl oz at V8 fb VT/R1	0.9	0.3	0.3	2.4 bc	1.2 b
Dexter Xcel 48.0 fl oz at V10	0.8	0.8	0.9	4.2 ab	1.8 b
Dexter Xcel 48.0 fl oz at VT/R1	0.8	0.6	0.7	2.6 bc	1.0 b
Brixen 13.0 fl oz at VT/R1	1.0	0.8	0.7	1.4 c	1.2 b
Fortix 5.0 fl oz at VT/R1	0.7	0.6	0.8	2.3 c	1.9 b
Delaro Complete 458 SC 8.0 fl oz at VT/R1	1.0	0.8	1.0	1.8 c	1.7 b
Veltyma 3.34 SC 7.0 fl oz at VT/R1	0.8	0.6	0.6	1.4 c	2.2 b
Quilt Xcel 2.2 SE 10.5.0 fl oz at VT/R1	0.8	0.8	0.9	2.6 bc	1.4 b
Headline AMP 1.68 SC 10 fl oz at VT/R1	0.8	0.3	0.6	2.1 c	1.3 b
P-value[w]	*0.9951*	*0.1283*	*0.2237*	*0.0012*	*0.0130*

[z] Fungicide treatments were applied on July 15 at V8, July 19 at V10, and August 7 at the tassel/silk (VT/R1) growth stages, and all treatments contained a nonionic surfactant (Preference) at a rate of 0.25% v/v. fb = followed by.

[y] Tar spot stroma was visually assessed as a percentage (0–100%) of leaf area on five plants in each plot at the ear leaf (EL), ear leaf minus two (EL–2), ear leaf plus two (EL+2) on September 21.

[x] Disease severity was visually assessed as a percentage (0–100%) of symptomatic leaf area on ear leaf on October 2. Five plants were assessed per plot and averaged before analysis. GLS = gray leaf spot, CR = Common rust.

[w] Means followed by the same letter are not significantly different based on Fisher's Least Significant Difference test (LSD; α=0.05).

TABLE 26. *Effect of Fungicide on Tar Spot on October 2*

TREATMENT, RATE/ACRE, AND TIMING[z]	TAR SPOT % STROMA[y] EL-2	TAR SPOT % STROMA[y] EL	TAR SPOT % STROMA[y] EL+2	TAR SPOT CHLO/ NECR[x] EL-2	TAR SPOT CHLO/ NECR[x] EL	TAR SPOT CHLO/ NECR[x] EL+2
Nontreated control	7.0	4.7 a	6.2	71.5 a	47.8 a	21.7 a
Dexter Xcel 48.0 fl oz at V8 fb VT/R1	3.4	2.5 cd	4.0	38.2 cd	6.8 cd	3.3 bc
Dexter Xcel 48.0 fl oz at V10	5.5	5.0 a	4.8	46.5 bcd	21.9 bcd	7.6 bc
Dexter Xcel 48.0 fl oz at VT/R1	3.2	3.2 a-d	4.1	50.5 a-d	14.3 bcd	1.5 c
Brixen 13.0 fl oz at VT/R1	2.7	2.6 bcd	3.8	48.8 a-d	12.4 bcd	2.8 bc
Fortix 5.0 fl oz at VT/R1	3.7	4.0 a-d	5.4	63.3 ab	23.9 bcd	8.3 bc
Delaro Complete 458 SC 8.0 fl oz at VT/R1	3.7	4.3 abc	5.6	58.0 a-c	22.5 bc	10.3 b
Veltyma 3.34 SC 7.0 fl oz at VT/R1	2.9	2.2 d	5.5	29.3 d	6.7 d	4.8 bc
Quilt Xcel 2.2 SE 10.5.0 fl oz at VT/R1	4.4	4.4 ab	5.8	52.3 a-d	20.8 bcd	6.5 bc
Headline AMP 1.68 SC 10 fl oz at VT/R1	2.4	3.5 a-d	5.3	59.0 abc	16.8 bcd	7.8 bc
P-value[w]	*0.1647*	*0.0386*	*0.1543*	*0.062*	*0.0009*	*0.0018*

[z] Fungicide treatments were applied on July 15 at V8, July 19 at V10, and August 7 at the tassel/silk (VT/R1) growth stages, and all treatments contained a nonionic surfactant (Preference) at a rate of 0.25% v/v. fb = followed by.

[y] Tar spot stroma was visually assessed as a percentage (0–100%) of leaf area on five plants in each plot at the ear leaf (EL), ear leaf minus two (EL-2), and ear leaf plus two (EL+2).

[x] Tar spot chlorosis and necrosis symptoms were visually assessed as a percentage (0–100%) of leaf area on five plants in each plot at the EL, EL-2, and EL+2.

[w] Means followed by the same letter are not significantly different based on Fisher's Least Significant Difference test (LSD; α=0.05).

TABLE 27. ***Effect of Fungicide on Canopy Greenness, Lodging, and Corn Yield***

TREATMENT, RATE/ACRE, AND TIMING[z]	CANOPY GREEN[y] %	LODGING[x] %	HARVEST MOISTURE %	TEST WEIGHT LB/BU	YIELD[w] BU/ACRE
Nontreated control	43.8 c	5.0	21.2 c	53.0	201.2
Dexter Xcel 48.0 fl oz at V8 fb VT/R1	78.8 a	0.0	22.2 ab	51.9	214.4
Dexter Xcel 48.0 fl oz at V10	70.0 ab	2.5	21.6 bc	52.2	200.6
Dexter Xcel 48.0 fl oz at VT/R1	73.8 a	0.0	22.0 ab	52.5	216.6
Brixen 13.0 fl oz at VT/R1	80.0 ab	0.0	22.5 a	52.1	213.2
Fortix 5.0 fl oz at VT/R1	73.8 b	0.0	21.4 bc	51.9	207.1
Delaro Complete 458 SC 8.0 fl oz at VT/R1	63.8 a	0.0	21.6 bc	52.6	211.3
Veltyma 3.34 SC 7.0 fl oz at VT/R1	78.8 ab	0.0	22.5 a	51.9	217.7
Quilt Xcel 2.2 SE 10.5.0 fl oz at VT/R1	72.5 ab	0.0	21.5 bc	52.3	217.5
Headline AMP 1.68 SC 10 fl oz at VT/R1	76.3 a	0.0	21.6 bc	52.0	209.7
P-value[v]	*0.0001*	*0.5507*	*0.0196*	*0.1263*	*0.5495*

[z] Fungicide treatments were applied on July 15 at V8, July 19 at V10, and August 7 at the tassel/silk (VT/R1) growth stages, and all treatments contained a nonionic surfactant (Preference) at a rate of 0.25% v/v. fb = followed by.

[y] Canopy greenness was visually assessed as a percentage (0–100%) of canopy October 2.

[x] Lodging = percentage of lodged stalks when pushed from shoulder height to 45° from vertical on October 2.

[w] Yields were adjusted to 15.5% moisture at harvest on October 28.

[v] Means followed by the same letter are not significantly different based on Fisher's Least Significant Difference test (LSD; α=0.05).

EVALUATION OF FUNGICIDES AND TIMING FOR TAR SPOT IN CORN IN NORTHWESTERN INDIANA, 2019 (COR19-15.PPAC)

C. Rocca da Silva, S. Shim, J. D. Ravellette, and D. E. P. Telenko, Department of Botany and Plant Pathology, Purdue University West Lafayette, Indiana 47907-2054

CORN (*ZEA MAYS* W2585SSRIB)

Tar spot, *Phyllachora maydis*

A trial was established at the Pinney Purdue Agricultural Center (PPAC) in Porter County, IN. The experiment was a randomized complete block design with four replications. Plots were 10 feet wide and 30 feet long and consisted of four rows, and the two center rows were used for evaluation. The previous crop was corn. Standard practices for grain corn production in Indiana were followed. Corn hybrid W2585SSRIB was planted in 30-inch row spacing at a rate of 34,000 seeds/acre on June 8. The field was overhead irrigated weekly at 1 inch unless weekly rainfall was 1 inch or higher to encourage disease. All fungicide applications were applied at 15 gal/acre and 40 psi using a Lee self-propelled sprayer equipped with a 10-foot boom, fitted with six TJ-VS 8002 nozzles spaced 20 inches apart, at 3.6 mph. Fungicides were applied on July 3 at the V5 growth stage and on August 9 at the R1 (silk) growth stage. Disease ratings were assessed on September 21 and October 2 at the dent (R5) and maturity (R6) growth stages, respectively. Tar spot was rated by visually assessing the percentage of stroma and the percentage of symptomatic tissues (chlorosis and necrosis) per leaf on five plants in each plot at the ear leaf (EL), ear leaf minus two (EL−2), and ear leaf plus two (EL+2). Values for each plot were averaged before analysis. The two center rows of each plot were harvested on October 25, and yields were adjusted to 15.5% moisture. Data were subjected to analysis of variance (SAS 9.4, 2019), and means were compared using Fisher's Least Significant Difference test (LSD; α=0.05).

In 2019, weather conditions were favorable for disease. Tar spot was the most prominent disease in the trial and reached moderate to high severity. Delaro 325 SC, Trivapro, and Miravis Neo application at the R1 or V5 plus R1 growth stage significantly reduced tar spot severity on the EL−2, EL, and EL+2 on September 21 (Table 28). Single fungicide applications at the V5 growth stage were not different from the nontreated control for tar spot on the EL−2 but significantly reduced tar spot on the EL and the EL+2 on September 21. The V5 applications were not significantly different from the nontreated control for tar spot stroma or leaf blight symptoms on all leaves on October 2 (Table 29). All fungicide treatments that included an R1 application significantly reduced tar spot stroma and chlorotic and necrotic symptoms on all leaves on October 2. There were no significant differences between these fungicide treatments except for Delaro 325 SC, which had significantly less tar spot stroma on the EL versus a single application of Trivapro or Miravis Neo at R1, and a double application of Delaro 325 SC showed less EL blight than a single application of Trivapro 2.21 SE. All fungicide treatments that included an R1 application significantly increased canopy greenness, reduced lodging, and increased yield over the nontreated control (Table 30).

TABLE 28. ***Effect of Fungicide Treatment on Tar Spot on September 21***

TREATMENT, RATE/ACRE, AND TIMING[Z]	TAR SPOT % STROMA[Y] EL−2	TAR SPOT % STROMA[Y] EL	TAR SPOT % STROMA[Y] EL+2
Nontreated control	7.0 a	4.0 a	1.9 a
Delaro 325 SC 8.0 fl oz at V5	5.8 a	2.5 b	1.4 b
Delaro 325 SC 8.0 fl oz at R1	1.3 b	1.0 c	1.0 b
Delaro 325 SC 8.0 fl oz at V5 fb R1	1.3 b	1.0 c	1.0 b
Trivapro 2.21 SE 13.7 fl oz at V5	5.4 a	2.8 b	1.3 b
Trivapro 2.21 SE 13.7 fl oz at R1	1.8 b	1.3 c	1.0 b
Trivapro 2.21 SE 13.7 fl oz at V5 fb R1	1.6 b	1.3 c	1.0 b
Miravis Neo 2.5 SE 13.7 fl oz at V5	7.6 a	2.3 b	1.4 b
Miravis Neo 2.5 SE 13.7 fl oz at R1	2.7 b	1.2 c	1.0 b
Miravis Neo 2.5 SE 13.7 fl oz at V5 fb R1	1.8 b	1.2 c	1.0 b
P-value[x]	*0.0001*	*0.0001*	*0.0030*

[z] Fungicide treatments were applied on July 3 at the V5 growth stage and on August 9 at the silk (R1) growth stage, and all treatments contained a nonionic surfactant (Preference) at a rate of 0.25% v/v. fb = followed by.

[y] Tar spot stroma was visually assessed as a percentage (0–100%) of leaf area on five plants in each plot at the ear leaf, ear leaf minus two, and ear leaf plus two.

[x] Means followed by the same letter are not significantly different based on Fisher's Least Significant Difference test (LSD; α=0.05).

TABLE 29. ***Effect of Fungicide Treatment on Tar Spot on October 2***

TREATMENT, RATE/ACRE, AND TIMING[z]	TAR SPOT % STROMA[Y] TAR SPOT EL–2	TAR SPOT % STROMA[Y] EL	TAR SPOT % STROMA[Y] EL+2	TAR SPOT % CHLO/ NECR[X] EL–2	TAR SPOT % CHLO/ NECR[X] EL	TAR SPOT % CHLO/ NECR[X] EL+2
Nontreated control	37.3 a	35.3 a	23.4 a	92.8 a	81.8 a	33.3 ab
Delaro 325 SC 8.0 fl oz at V5	35.0 a	30.0 b	20.7 a	88.0 a	72.3 a	30.0 b
Delaro 325 SC 8.0 fl oz at R1	20.1 cd	10.9 e	8.3 b	58.3 b	15.6 bc	6.6 c
Delaro 325 SC 8.0 fl oz at V5 fb R1	19.3 d	10.8 e	6.7 b	58.0 b	12.4 c	7.3 c
Trivapro 2.21 SE 13.7 fl oz at V5	37.3 a	32.8 ab	24.7 a	95.3 a	77.5 a	47.3 a
Trivapro 2.21 SE 13.7 fl oz at R1	26.8 b	18.5 c	9.4 b	66.8 b	24.8 b	7.9 c
Trivapro 2.21 SE 13.7 fl oz at V5 fb R1	27.8 b	14.7 cde	8.9 b	66.3 b	23.1 bc	9.1 c
Miravis Neo 2.5 SE 13.7 fl oz at V5	37.8 a	35.0 a	22.8 a	91.3 a	79.0 a	29.5 b
Miravis Neo 2.5 SE 13.7 fl oz at R1	22.9 bcd	15.0 cd	7.4 b	59.0 b	22.9 bc	6.7 c
Miravis Neo 2.5 SE 13.7 fl oz at V5 fb R1	25.0 bc	12.5 de	7.8 b	58.0 b	14.8 bc	7.8 c
P-value[w]	*0.0001*	*0.0001*	*0.0001*	*0.0001*	*0.0001*	*0.0001*

[z] Fungicide treatments were applied on July 3 at the V5 growth stage and on August 9 at the R1 (silk) growth stage, and all treatments contained a nonionic surfactant (Preference) at a rate of 0.25% v/v. fb = followed by.

[y] Tar spot stroma was visually assessed as a percentage (0–100%) of leaf area on five plants in each plot at the ear leaf (EL), ear leaf minus two (EL–2), and ear leaf plus two (EL+2).

[x] Tar spot chlorosis and necrosis symptoms were visually assessed as a percentage (0–100%) of leaf area on five plants in each plot at the EL, EL–2, and EL+2.

[w] Means followed by the same letter are not significantly different based on Fisher's Least Significant Difference test (LSD; α=0.05).

TABLE 30. ***Effect of Fungicide Treatment on Canopy Greenness, Lodging, and Corn Yield***

TREATMENT, RATE/ACRE, AND TIMING[z]	CANOPY GREEN[y] % SEPTEMBER 21	CANOPY GREEN[y] % OCTOBER 2	LODGING[x] %	HARVEST MOISTURE %	TEST WEIGHT BU/ACRE	YIELD[w] LB/BU
Nontreated control	80.0 c	35.0 c	2.0 a	23.2 b	51.2	201.6 d
Delaro 325 SC 8.0 fl oz at V5	83.8 bc	47.5 b	0.5 cd	23.1 b	50.5	206.9 bcd
Delaro 325 SC 8.0 fl oz at R1	87.5 ab	68.8 a	0.5 cd	24.7 a	50.8	221.7 a
Delaro 325 SC 8.0 fl oz at V5 fb R1	91.3 a	76.3 a	0.8 bcd	24.9 a	50.8	223.4 a
Trivapro 2.21 SE 13.7 fl oz at V5	83.8 bc	36.3 c	1.3 abc	23.2 b	50.8	204.4 cd
Trivapro 2.21 SE 13.7 fl oz at R1	91.3 a	66.3 a	0.0 d	24.6 a	51.6	215.5 abc
Trivapro 2.21 SE 13.7 fl oz at V5 fb R1	91.3 a	67.5 a	0.3 d	24.3 a	50.8	216.9 ab
Miravis Neo 2.5 SE 13.7 fl oz at V5	83.8 bc	42.5 bc	1.5 ab	23.4 b	51.0	205.3 bcd
Miravis Neo 2.5 SE 13.7 fl oz at R1	90.0 a	71.3 a	0.0 d	24.3 a	50.8	212.0 a
Miravis Neo 2.5 SE 13.7 fl oz at V5 fb R1	90.0 a	72.5 a	0.0 d	24.6 a	50.6	225.0 a
P-value[v]	*0.0031*	*0.0001*	*0.0012*	*0.0001*	*0.8632*	*0.0009*

[z] Fungicide treatments were applied on July 3 at the V5 growth stage and on August 9 at the R1 (silk) growth stage, and all treatments contained a nonionic surfactant (Preference) at a rate of 0.25% v/v. fb = followed by.

[y] Canopy greenness was visually assessed as a percentage (0–100%) of canopy on September 21 and October 2.

[x] Lodging = percentage of lodged stalks when pushed from shoulder height to 45° from vertical.

[w] Yields were adjusted to 15.5% moisture at harvest on October 25.

[v] Means followed by the same letter are not significantly different based on Fisher's Least Significant Difference test (LSD; α=0.05).

FUNGICIDE EVALUATION FOR FOLIAR DISEASE IN IRRIGATED CORN IN NORTHWESTERN INDIANA, 2019 (COR19-16.PPAC)

S. Shim, J. D. Ravellette, and D. E. P. Telenko, Department of Botany and Plant Pathology, Purdue University West Lafayette, Indiana 47907-2054

CORN (*ZEA MAYS* W2585SSRIB)

Tar spot, *Phyllachora maydis*

A trial was established at the Pinney Purdue Agricultural Center (PPAC) in Porter County, IN. The experiment was a randomized complete block design with four replications. Plots were 10 feet wide and 30 feet long and consisted of four rows, and the two center rows were used for evaluation. The previous crop was corn. Standard practices for grain corn production in Indiana were followed. Corn hybrid W2585SSRIB was planted in 30-inch row spacing at a rate of 34,000 seeds/acre on June 8. The field was overhead irrigated weekly at 1 inch unless weekly rainfall was 1 inch or higher to encourage disease. All fungicide applications were applied at 15 gal/acre and 40 psi using a Lee self-propelled sprayer equipped with a 10-foot boom, fitted with six TJ-VS 8002 nozzles spaced 20 inches apart, at 3.6 mph. Fungicides were applied on July 19 at the V10 growth stage and on August 8 at the tassel/silk (VT/R1) growth stage. Disease ratings were assessed on September 21 and October 2 at the dent (R5) and maturity (R6) growth stages, respectively. Tar spot was rated by visually assessing the percentage of stroma and the percentage of symptomatic tissues (chlorosis and necrosis) per leaf on five plants in each plot at the ear leaf, ear leaf minus two, and ear leaf plus two. Values for each plot were averaged before analysis. The two center rows of each plot were harvested on October 28, and yields were adjusted to 15.5% moisture. Data were subjected to analysis of variance (SAS 9.4, 2019), and means were compared using Fisher's Least Significant Difference test (LSD; α=0.05).

In 2019, weather conditions were favorable for disease. Tar spot was the most prominent diseases in the trial and reached moderate to high severity. All fungicide treatments reduced severity of tar spot stroma over the nontreated control at all rating dates (Tables 31 and 32). The percent of chlorotic and necrotic symptoms were also significantly reduced with all fungicide treatments on October 2 (Table 32). All fungicide treatments significantly increased the percentage of canopy greenness over the nontreated control on both September 21 and October 2 (Table 33). All fungicide treatments significantly reduced lodging and increased and yield over the nontreated control. Corn yield was greatest in plots treated with Veltyma 9.0 fl oz but was not significantly different from fungicide treatments of Delaro 325 SC, Trivapro, Headline AMP at 10.0 fl oz, Veltyma at 7.0 fl oz, and Priaxor Xemium followed by Veltyma.

TABLE 31. ***Effect of Fungicide on Tar Spot on September 21***

TREATMENT, RATE/ACRE, AND TIMING[z]	TAR SPOT % STROMA[Y] EL–2	TAR SPOT % STROMA[Y] EL	TAR SPOT % STROMA[Y] EL+2
Nontreated control	21.2 a	5.6 a	2.7 a
Lucento 4.17 SC 5.0 fl oz at VT	2.8 b	1.4 b	1.3 b
Delaro 325 SC 8.0 fl oz at VT	1.7 b	1.4 b	1.1 c
Trivapro 2.21 SE 13.7 fl oz at VT	3.3 b	1.6 b	1.3 b
Miravis Neo 2.5 SE 13.7 fl oz at VT	4.2 b	1.4 b	1.0 c
Headline AMP 1.68 SC 10.0 fl oz at VT	2.3 b	1.2 b	1.2 bc
Veltyma 3.34 SC 7.0 fl oz at VT	1.8 b	1.1 b	1.0 c
Priaxor Xemium 4.0 fl oz at V10 fb Veltyma 3.34 SC 7.0 fl oz at VT	1.3 b	1.2 b	1.0 c
Veltyma 3.34 SC 9.0 fl oz at VT	1.4 b	1.1 b	1.0 c
Headline AMP 1.68 SC 14.4 fl oz at VT	2.4 b	1.3 b	1.1 bc
P-value[x]	*0.0001*	*0.0001*	*0.0001*

[z] Fungicide treatments were applied on July 19 at the V10 growth stage and on August 8 at the tassel/silk (VT/R1) growth stage, and all treatments contained a nonionic surfactant (Preference) at a rate of 0.25% v/v. fb = followed by.

[y] Tar spot stroma was visually assessed as a percentage (0–100%) of leaf area on five plants in each plot at the ear leaf, ear leaf minus two, and ear leaf plus two.

[x] Means followed by the same letter are not significantly different based on Fisher's Least Significant Difference test (LSD; α=0.05).

TABLE 32. ***Effect of Fungicide on Tar Spot on October 2***

TREATMENT, RATE/ACRE, AND TIMING[z]	TAR SPOT % STROMA[y] EL-2	TAR SPOT % STROMA[y] EL	TAR SPOT % STROMA[y] EL+2	TAR SPOT % CHLOR/ NEC[x] EL-2	TAR SPOT % CHLOR/ NEC[x] EL	TAR SPOT % CHLOR/ NEC[x] EL+2
Nontreated control	40.8 a	40.8 a	37.1 a	98.8 a	94.8 a	89.2 a
Lucento 4.17 SC 5.0 fl oz at VT	27.8 bcd	23.8 b	15.2 bc	75.5 b-e	59.5 b	22.5 b
Delaro 325 SC 8.0 fl oz at VT	28.0 bcd	20.3 bc	12.4 bcd	79.0 bcd	34.5 cd	15.0 b-e
Trivapro 2.21 SE 13.7 fl oz at VT	33.3 ab	24.3 b	13.7 bc	86.3 ab	54.8 b	16.3 bcd
Miravis Neo 2.5 SE 13.7 fl oz at VT	32.8 bc	20.8 bc	10.6 cde	83.5 abc	46.0 b	15.2 b-e
Headline AMP 1.68 SC 10.0 fl oz at VT	25.3 cd	14.7 d	16.0 b	69.5 b-e	28.1 d	17.6 bc
Veltyma 3.34 SC 7.0 fl oz at VT	24.3 d	12.0 d	10.4 cde	68.3 b-e	12.2 ef	7.4 cde
Priaxor Xemium 4.0 fl oz at V10 fb Veltyma 3.34 SC 7.0 fl oz at VT	13.0 f	6.4 e	7.0 e	60.0 de	9.7 f	5.0 de
Veltyma 3.34 SC 9.0 fl oz at VT	16.3 ef	10.9 de	7.8 de	56.5 e	10.6 f	3.8 e
Headline AMP 1.68 SC 14.4 fl oz at VT	21.4 de	16.0 cd	10.2 cde	64.8 cde	26.7 de	21.0 b
P-value[w]	*0.0001*	*0.0001*	*0.0001*	*0.0029*	*0.0001*	*0.0001*

[z] Fungicide treatments were applied on July 19 at the V10 growth stage and on August 8 at the tassel/silk (VT/R1) growth stage, and all treatments contained a nonionic surfactant (Preference) at a rate of 0.25% v/v. fb = followed by.

[y] Tar spot stroma was visually assessed as a percentage (0–100%) of leaf area on five plants in each plot at the ear leaf (EL), ear leaf minus two (EL-2), and ear leaf plus two (EL+2).

[x] Tar spot chlorosis and necrosis symptoms were visually assessed as a percentage (0–100%) of leaf area on five plants in each plot at the EL, EL-2, and ear leaf plus two.

[w] Means followed by the same letter are not significantly different based on Fisher's Least Significant Difference test (LSD; α=0.05).

TABLE 33. ***Effect of Fungicide on Canopy Greenness, Lodging, and Yield***

TREATMENT, RATE/ACRE, AND TIMING[z]	CANOPY GREEN[y] % SEPTEMBER 21	CANOPY GREEN[y] % OCTOBER 2	LODGING[x] %	HARVEST MOISTURE %	TEST WEIGHT LB/BU	YIELD[w] BU/ACRE
Nontreated control	61.3 b	21.3 d	40.0 a	23.2 d	50.7	179.8 e
Lucento 4.17 SC 5.0 fl oz at VT	92.5 a	45.0 c	25.0 b	23.6 bcd	49.8	197.1 d
Delaro 325 SC 8.0 fl oz at VT	93.8 a	58.8 bc	0.0 b	23.4 cd	50.4	209.6 a-d
Trivapro 2.21 SE 13.7 fl oz at VT	91.3 a	45.0 c	0.0 b	23.3 cd	53.6	209.3 a-d
Miravis Neo 2.5 SE 13.7 fl oz at VT	88.8 a	47.5 c	25.0 b	23.3 cd	50.5	202.8 bcd
Headline AMP 1.68 SC 10.0 fl oz at VT	92.5 a	65.0 ab	25.0 b	23.6 cd	50.4	211.2 abc
Veltyma 3.34 SC 7.0 fl oz at VT	92.5 a	68.8 ab	25.0 b	24.4 ab	50.5	214.5 ab
Priaxor Xemium 4.0 fl oz at V10 fb Veltyma 3.34 SC 7.0 fl oz at VT	95.0 a	75.0 a	0.0 b	25.0 a	50.3	213.7 ab
Veltyma 3.34 SC 9.0 fl oz at VT	95.0 a	73.8 a	25.0 b	25.1 a	50.2	223.5 a
Headline AMP 1.68 SC 14.4 fl oz at VT	91.3 a	58.8 bc	0.0 b	24.1 bc	50.5	193.4 d
P-value[v]	*0.0001*	*0.0001*	*0.0001*	*0.0001*	*0.3575*	0.0002

[z] Fungicide treatments were applied on July 19 at the V10 growth stage and on August 8 at the tassel/silk (VT/R1) growth stage, and all treatments contained a nonionic surfactant (Preference) at a rate of 0.25% v/v. fb = followed by.

[y] Canopy greenness was visually assessed as a percentage (0–100%) of canopy on September 21 and October 2.

[x] Lodging = percentage of lodged stalks when pushed from shoulder height to 45° from vertical on October 2.

[w] Yields were adjusted to 15.5% moisture at harvest on October 28.

[v] Means followed by the same letter are not significantly different based on Fisher's Least Significant Difference test (LSD; α=0.05).

ASSESSMENT OF FUNGICIDES APPLIED AT TASSEL/SILK (VT/R1) FOR TAR SPOT IN CORN IN NORTHWESTERN INDIANA, 2019 (COR19-17.PPAC)

S. Shim, J. D. Ravellette, and D. E. P. Telenko, Department of Botany and Plant Pathology, Purdue University West Lafayette, Indiana 47907-2054

CORN (*ZEA MAYS* W2585SSRIB)

Tar spot, *Phyllachora maydis*

A trial was established at the Pinney Purdue Agricultural Center (PPAC) in Porter County, IN. The experiment was a randomized complete block design with four replications. Plots were 10 feet wide and 30 feet long and consisted of four rows, and the two center rows were used for evaluation. The previous crop was corn. Standard practices for grain corn production in Indiana were followed. Corn hybrid W2585SSRIB was planted in 30-inch row spacing at a rate of 34,000 seeds/acre on June 8. The field was overhead irrigated weekly at 1 inch unless weekly rainfall was 1 inch or higher to encourage disease. All fungicide applications were applied at 15 gal/acre and 40 psi using a Lee self-propelled sprayer equipped with a 10-foot boom, fitted with six TJ-VS 8002 nozzles spaced 20 inches apart, at 3.6 mph. Fungicides were applied on August 8 at the tassel/silk (VT/R1) growth stage. Disease ratings were assessed on September 21 and October 2 at the dent (R5) and maturity (R6) growth stages, respectively. Tar spot was rated by visually assessing the percentage of stroma and the percentage of symptomatic tissues (chlorosis and necrosis) per leaf on five plants in each plot at the ear leaf, ear leaf minus two, and ear leaf plus two. Values for each plot were averaged before analysis. The two center rows of each plot were harvested on October 25, and yields were adjusted to 15.5% moisture. Data were subjected to analysis of variance, and means were compared using Fisher's Least Significant Difference (LSD; α=0.05).

In 2019, weather conditions were favorable for disease. Tar spot was the most prominent disease in the trial and reached moderate to high severity. All fungicide treatments reduced severity of tar spot stroma over the nontreated control at all rating dates (Tables 34 and 35). The percentage of chlorosis and necrosis symptoms were also significantly reduced with all fungicide treatments on October 2 (Table 35). All fungicide treatments significantly increased the percentage of canopy greenness of the corn over the nontreated control on both September 21 and October 2 (Table 36). Treatments of Priaxor Xemium, Quilt Xcel, and Miravis Neo were less green than Veltyma, Revytek, and Headline AMP. All fungicide treatments significantly reduced lodging and increased yield over the nontreated control. Yield was highest with Veltyma 3.34 SC, Revytek 3.33 LC, Stratego YLD, and Miravis Neo 2.5 SE.

TABLE 34. ***Effect of Fungicide on Tar Spot on September 21***

TREATMENT AND RATE/ACRE[z]	TAR SPOT % STROMA[y] EL–2	TAR SPOT % STROMA[y] EL	TAR SPOT % STROMA[y] EL+2
Nontreated control	14.3 a	4.8 a	2.3 a
Veltyma 3.34 SC 7.0 fl oz	1.2 b	1.1 b	1.0 b
Revytek 3.33 LC 8.0 fl oz	2.1 b	1.2 b	1.0 b
Trivapro 2.21 SE 13.7 fl oz	2.7 b	1.4 b	1.0 b
Headline AMP 1.68 SC 10.0 fl oz	1.5 b	1.1 b	1.2 b
Priaxor Xemium 4.0 fl oz	2.1 b	1.2 b	1.1 b
Quilt Xcel 2.2 SE 14.5 fl oz	2.0 b	1.0 b	1.1 b
Stratego YLD 4.0 fl oz	1.9 b	1.1 b	1.1 b
Miravis Neo 2.5 SE 13.7 fl oz	2.4 b	1.2 b	1.0 b
P-value[x]	0.0005	0.0001	0.0001

[z] Fungicide treatments were applied on August 8 at the tassel/silk (VT/R1) growth stage, and all treatments contained a nonionic surfactant (Preference) at a rate of 0.25% v/v.

[y] Tar spot stroma was visually assessed as a percentage (0–100%) of leaf area on five plants in each plot at the ear leaf, ear leaf minus two, and ear leaf plus two on September 21.

[x] Means followed by the same letter are not significantly different based on Fisher's Least Significant Difference test (LSD; α=0.05).

TABLE 35. ***Effect of Fungicide on Tar Spot on October 2***

TREATMENT AND RATE/ACRE[z]	TAR SPOT % STROMA[Y] EL-2	TAR SPOT % STROMA[Y] EL	TAR SPOT % STROMA[Y] EL+2	TAR SPOT CHLOR/ NECR[X] EL-2	TAR SPOT CHLOR/ NECR[X] EL	TAR SPOT % CHLOR/ NECR[X] EL+2
Nontreated control	38.3 a	37.3 a	26.3 a	98.0 a	83.5 a	82.0 a
Veltyma 3.34 SC 7.0 fl oz	12.0 f	6.9 g	5.7 c	44.5 e	5.4 f	2.9 b
Revytek 3.33 LC 8.0 fl oz	15.2 ef	9.0 fg	6.1 c	45.8 de	10.3 ef	9.6 b
Trivapro 2.21 SE 13.7 fl oz	28.3 b	18.6 b	9.4 bc	67.8 b	34.8 b	11.0 b
Headline AMP 1.68 SC 10.0 fl oz	15.7 def	10.7 ef	8.3 bc	49.5 cde	15.4 b	8.4 b
Priaxor Xemium 4.0 fl oz	24.0 bc	16.1 bc	9.5 bc	57.5 b-e	25.0 de	8.5 b
Quilt Xcel 2.2 SE 14.5 fl oz	21.8 cd	14.6 cd	6.8 bc	60.0 bcd	19.6 c	9.0 b
Stratego YLD 4.0 fl oz	21.0 cde	10.9 ef	10.7 b	52.0 cde	17.7 cd	9.9 b
Miravis Neo 2.5 SE 13.7 fl oz	21.3 cde	12.9 de	6.5 bc	63.3 bc	22.4 cde	3.9 b
P-value[w]	*0.0001*	*0.0001*	*0.0001*	*0.0001*	*0.0001*	*0.0001*

[z] Fungicide treatments were applied on August 8 at the VT/R1 growth (tassel/silk) stage, and all treatments contained a nonionic surfactant (Preference) at a rate of 0.25% v/v.

[y] Tar spot stroma was visually assessed as a percentage (0–100%) of leaf area on five plants in each plot at the ear leaf (EL), ear leaf minus two (EL-2), and ear leaf plus two (EL+2).

[x] Tar spot chlorosis and necrosis symptoms were visually assessed as a percentage (0–100%) of leaf area on five plants in each plot at the EL, EL-2, and EL+2.

[w] Means followed by the same letter are not significantly different based on Fisher's Least Significant Difference test (LSD; α=0.05).

TABLE 36. ***Effect of Fungicide on Canopy Greenness, Lodging, and Yield***

TREATMENT AND RATE/ACRE[z]	CANOPY GREEN[y] %	CANOPY GREEN[y] %	LODGING[x] %	HARVEST MOISTURE %	TEST WEIGHT LB/BU	YIELD[w] BU/ACRE
Nontreated control	75.0 b	28.8 c	28.0 a	23.1 d	50.1	198.3 e
Veltyma 3.34 SC 7.0 fl oz	95.0 a	78.8 a	0.0 b	25.7 a	50.8	229.1 a
Revytek 3.33 LC 8.0 fl oz	91.3 a	78.8 a	0.0 b	25.0 ab	50.7	225.7 abc
Trivapro 2.21 SE 13.7 fl oz	93.8 a	70.0 ab	3.0 b	24.5 bc	50.2	216.0 d
Headline AMP 1.68 SC 10.0 fl oz	95.0 a	78.8 a	0.0 b	24.6 bc	49.5	217.0 cd
Priaxor Xemium 4.0 fl oz	95.0 a	66.3 b	3.0 b	24.2 c	50.3	215.9 d
Quilt Xcel 2.2 SE 14.5 fl oz	93.8 a	65.0 b	0.0 b	24.2 c	50.4	219.6 bcd
Stratego YLD 4.0 fl oz	93.8 a	71.3 ab	0.0 b	24.1 c	50.5	220.0 a-d
Miravis Neo 2.5 SE 13.7 fl oz	95.0 a	62.5 b	0.0 b	24.3 c	50.6	228.3 ab
P-value[v]	*0.0001*	*0.0001*	*0.0001*	*0.0001*	*0.1925*	*0.0001*

[z] Fungicide treatments were applied on August 8 at the VT/R1 growth (tassel/silk) stage, and all treatments contained a nonionic surfactant (Preference) at a rate of 0.25% v/v.

[y] Canopy greenness was visually assessed as a percentage (0–100%) of canopy on September 21 and October 2.

[x] Lodging = percentage of lodged stalks when pushed from shoulder height to 45° from vertical.

[w] Yields were adjusted to 15.5% moisture at harvest on October 28

[v] Means followed by the same letter are not significantly different based on Fisher's Least Significant Difference test (LSD; α=0.05).

FUNGICIDE EVALUATION FOR FOLIAR DISEASE IN NONIRRIGATED CORN IN NORTHWESTERN INDIANA, 2019 (COR19-22.PPAC)

C. Rocco da Silva, S. Shim, J. D. Ravellette, and D. E. P. Telenko, Department of Botany and Plant Pathology, Purdue University West Lafayette, Indiana 47907-2054

CORN (*ZEA MAYS* W2585SSRIB)

Tar spot, *Phyllachora maydis*

A trial was established at the Pinney Purdue Agricultural Center (PPAC) in Porter County, IN. The experiment was a randomized complete block design with four replications. Plots were 10 feet wide and 30 feet long and consisted of four rows, and the two center rows were used for evaluation. The previous crop was corn. Standard practices for nonirrigated grain corn production in Indiana were followed. Corn hybrid W2585SS-RIB was planted in 30-inch row spacing at a rate of 34,000 seeds/acre on June 8. All fungicide applications were applied at 15 gal/acre and 40 psi using a Lee self-propelled sprayer equipped with a 10-foot boom, fitted with six TJ-VS 8002 nozzles spaced 20 inches apart, at 3.6 mph. Fungicides were applied on July 3 at the V5 growth stage and on August 9 at the silk (R1) growth stage. Disease ratings were assessed on August 29, September 21 and September 30 at the milk (R3), dent (R5), and maturity (R6) growth stages, respectively. Disease severity was rated by visually assessing the percentage of symptomatic leaf area on five plants in each plot at the ear leaf, ear leaf minus two, and ear leaf plus two. Values for each plot were averaged before analysis. The two center rows of each plot were harvested on October 28, and yields were adjusted to 15.5% moisture. Data were subjected to analysis of variance (SAS 9.4, 2019), and means were compared using Fisher's Least Significant Difference test (LSD; α=0.05).

In 2019, weather conditions were favorable for disease. Tar spot was the most prominent disease in the trial and reached moderate to high severity. Fungicide applications of Affiance and Domark applied at R1 reduced tar spot stroma and chlorosis and necrosis symptoms for all rating dates (Tables 37 and 38). Fungicides applied at V5 generally were not significantly different from the nontreated control (Table 37 and 38). Affiance and Domark applied at R1 significantly increased the percentage of canopy greenness of the corn over the nontreated control on both August 21 and September 30 and increased yield over the nontreated control (Table 39).

TABLE 37. ***Effect of Fungicide on Tar Spot on September 21***

TREATMENT, RATE/ACRE, AND TIMING[z]	TAR SPOT % STROMA[y] EL–2	TAR SPOT % STROMA[y] EL	TAR SPOT % STROMA[y] EL+2	TAR SPOT % CHLOR/NECR[x] EL–2	TAR SPOT % CHLOR/NECR[x] EL	TAR SPOT % CHLOR/NECR[x] EL+2
Nontreated control	27.4 ab	7.4 a	3.4 a	46.5 a	7.0	0.7
Affiance 10.0 fl oz at V5	27.3 ab	7.1 a	2.2 bc	50.3 a	7.3	0.8
Domark 6.0 fl oz at V5	21.8 b	6.6 a	2.1 bc	44.8 a	4.9	0.4
Delaro 325 SC 8.0 fl oz at V5	27.5 ab	10.4 a	2.8 ab	52.3 a	10.9	2.8
Trivapro 2.21 SE 13.7 fl oz at V5	32.5 a	8.9 a	3.5 a	50.3 a	7.0	0.9
Affiance 10.0 fl oz at R1	12.0 c	2.1 b	1.5 cd	18.8 b	0.5	1.7
Domark 6.0 fl oz at R1	6.2 c	1.9 b	1.1 d	11.4 b	1.4	0.1
P-value[w]	*0.0001*	*0.0034*	*0.0003*	*0.0001*	*0.0575*	*0.1965*

[z] Fungicide treatments were applied on July 3 at the V5 growth stage and on August 9 at the silk (R1) growth stage.

[y] Tar spot stroma was visually assessed as a percentage (0–100%) of leaf area on five plants in each plot at the ear leaf (EL), ear leaf minus two (EL–2), ear leaf plus two (EL+2).

[x] Tar spot chlorosis and necrosis symptoms were visually assessed as a percentage (0–100%) of leaf area on five plants in each plot at the EL, EL–2, and EL+2.

[w] Means followed by the same letter are not significantly different based on Fisher's Least Significant Difference test (LSD; α=0.05).

TABLE 38. *Effect of Fungicide on Tar Spot on September 30*

TREATMENT, RATE/ACRE, AND TIMING[z]	TAR SPOT % STROMA[y] EL -2	TAR SPOT % STROMA[y] EL	TAR SPOT % STROMA[y] EL +2	TAR SPOT % CHLOR/ NECR[x] EL -2	TAR SPOT % CHLOR/ NECR[x] EL	TAR SPOT % CHLOR/ NECR[x] EL +2
Nontreated control	45.0 a	39.8 a	39.0 a	93.5	81.8 a	65.8 a
Affiance 10.0 fl oz at V5	40.8 ab	39.0 a	34.8 a	98.0	89.3 a	62.0 a
Domark 6.0 fl oz at V5	38.8 abc	39.0 a	34.8 a	98.5	83.3 a	54.0 a
Delaro 325 SC 8.0 fl oz at V5	36.3 bc	33.3 ab	30.3 a	80.8	65.6 a	47.6 a
Trivapro 2.21 SE 13.7 fl oz at V5	39.0 abc	38.5 a	32.3 a	96.5	86.5 a	52.8 a
Affiance 10.0 fl oz at R1	32.8 bc	27.8 b	17.0 b	73.3	32.8 b	10.7 b
Domark 6.0 fl oz at R1	31.5 c	25.3 b	19.3 b	69.5	32.0 b	14.3 b
P-value[w]	*0.0372*	*0.0197*	*0.0008*	*0.0882*	*0.0019*	*0.0001*

[z] Fungicide treatments were applied on July 3 at the V5 growth stage and on August 9 at the silk (R1) growth stage.

[y] Tar spot stroma was visually assessed as a percentage (0–100%) of leaf area on five plants in each plot at the ear leaf (EL), ear leaf minus two (EL–2), and ear leaf plus two (EL+2).

[x] Tar spot chlorosis and necrosis symptoms were visually assessed as a percentage (0–100%) of leaf area on five plants in each plot at the EL, EL–2, and EL+2.

[w] Means followed by the same letter are not significantly different based on Fisher's Least Significant Difference test (LSD; α=0.05).

TABLE 39. ***Effect of Fungicide on Canopy Greenness, Lodging, and Corn Yield***

TREATMENT, RATE/ACRE, AND TIMING[z]	CANOPY GREEN[y] % SEPTEMBER 21	CANOPY GREEN[y] % SEPTEMBER 30	LODGING[x] %	HARVEST MOISTURE %	TEST WEIGHT LB/BU	YIELD[w] BU/ACRE
Nontreated control	51.3 b	28.8 c	1.5	22.8	53.2	182.7 b
Affiance 10.0 fl oz at V5	61.3 b	33.8 bc	1.0	22.6	53.7	183.3 b
Domark 6.0 fl oz at V5	58.8 b	35.0 bc	1.3	22.0	53.2	176.7 b
Delaro 325 SC 8.0 fl oz at V5	63.8 b	32.5 bc	0.5	22.6	53.3	178.4 b
Trivapro 2.21 SE 13.7 fl oz at V5	63.8 b	36.3 b	0.8	22.0	53.4	181.5 b
Affiance 10.0 fl oz at R1	83.8 a	57.5 a	0.0	23.3	52.1	203.3 a
Domark 6.0 fl oz at R1	82.5 a	56.3 a	0.8	23.2	52.9	198.3 a
P-value[v]	*0.0001*	*0.0001*	*0.2399*	*0.1792*	*0.3662*	*0.0003*

[z] Fungicide treatments were applied on July 3 at the V5 growth stage and on August 9 at the silk (R1) growth stage.
[y] Canopy greenness was visually assessed as a percentage (0–100%) of canopy on September 21 and September 30.
[x] Lodging = percentage of lodged stalks when pushed from shoulder height to 45° from vertical on September 30.
[w] Yields were adjusted to 15.5% moisture at harvest on October 28.
[v] Means followed by the same letter are not significantly different based on Fisher's Least Significant Difference test (LSD; α=0.05).

EVALUATION OF A FUNGICIDE PROGRAMS FOR TAR SPOT IN CORN IN NORTHWESTERN INDIANA, 2019 (COR19-23.PPAC)

C. Rocca da Silva, S. Shim, J. D. Ravellette, and D. E. P. Telenko, Department of Botany and Plant Pathology, Purdue University West Lafayette, Indiana 47907-2054

CORN (*ZEA MAYS* W2585SSRIB)

Tar spot, *Phyllachora maydis*

A trial was established at the Pinney Purdue Agricultural Center (PPAC) in Porter County, IN. The experiment was a randomized complete block design with four replications. Plots were 10 feet wide and 30 feet long and consisted of four rows, and the two center rows were used for evaluation. The previous crop was corn. Standard practices for nonirrigated grain corn production in Indiana were followed. Corn hybrid W2585SS-RIB was planted in 30-inch row spacing at a rate of 34,000 seeds/acre on June 8. All fungicide applications were applied at 15 gal/acre and 40 psi using a Lee self-propelled sprayer equipped with a 10-foot boom, fitted with six TJ-VS 8002 nozzles spaced 20 inches apart, at 3.6 mph. Fungicides were applied on August 8 at the tassel/silk (VT/R1) and on August 22 at the blister (R2) growth stages. Disease ratings were assessed on September 21 and September 30 at the dent (R5) and maturity (R6) growth stages, respectively. Tar spot was rated by visually assessing the percentage of stroma and the percentage of symptomatic tissues (chlorosis and necrosis) per leaf on five plants in each plot at the ear leaf, ear leaf minus two, and ear leaf plus two. Values for each plot were averaged before analysis. The two center rows of each plot were harvested on October 25, and yields were adjusted to 15.5% moisture. Data were subjected to analysis of variance (SAS 9.4, 2019), and means were compared using Fisher's Least Significant Difference test (LSD; α=0.05).

In 2019, weather conditions were favorable for disease. Tar spot was the most prominent diseases and was first noted on August 21 and reached moderate to high severity. All fungicide treatments reduced severity of tar spot stroma over the nontreated control at all rating dates and leaf levels (Tables 40 and 41). Aproach Prima, Aproach followed by (fb) Aproach Prima, and Aproach Prima fb Aproach significantly reduced tar spot stroma severity over the nontreated control on September 30 (Tables 41). The percent of chlorosis and necrosis was also significantly reduced with all fungicide programs on September 21 on the ear leaf minus, two the ear leaf, and all leaves on September 30. All fungicide programs significantly increased the percent of canopy greenness over the nontreated control on both August 21 and September 30 (Table 42). No difference between treatments was detected for lodging. All fungicides treatments significantly increased yield over the nontreated control. Aproach Prima fb Aproach had the greatest yield but was not significantly different from the Aproach fb Aproach Prima program.

TABLE 40. *Effect of Fungicide on Tar Spot on September 21*

TREATMENTS, RATE/ACRE, AND TIMING[Z]	TAR SPOT % STROMA[Y] EL–2	TAR SPOT % STROMA[Y] EL	TAR SPOT % STROMA[Y] EL+2	TAR SPOT % CHLOR/ NECR[X] EL–2	TAR SPOT % CHLOR/ NECR[X] EL	TAR SPOT % CHLOR/ NECR[X] EL+2
Nontreated control	25.0 a	7.8 a	2.6 a	39.5 a	7.5 a	0.5
Aproach 6.0 fl oz at VT	10.8 b	2.7 b	1.7 b	18.8 b	2.2 b	1.0
Aproach Prima 2.34 SC 6.8 fl oz at VT	2.8 c	1.5 b	1.0 c	4.0 c	0.3 b	0.1
Aproach 6.0 fl oz at VT/R1 fb						
Aproach Prima 2.34 SC 6.8 fl oz R2	1.6 c	1.0 b	1.0 c	6.7 c	0.3 b	0.0
Aproach Prima 2.34 SC 6.8 fl oz at VT/R1 fb Aproach 6.0 fl oz R2	1.9 c	1.1 b	1.0 c	7.6 c	0.5 b	0.0
P-value[w]	*0.0001*	*0.0017*	*0.0003*	*0.0001*	*0.0093*	*0.2821*

[z] Fungicide treatments were applied on August 8 at the tassel/silk (VT/R1) and on August 22 at the blister (R2) growth stages, and all treatments contained a nonionic surfactant (Preference) at a rate of 0.25% v/v. fb = followed by.

[y] Tar spot stroma was visually assessed as a percentage (0–100%) of leaf area on five plants in each plot at the ear leaf (EL), ear leaf minus two (EL–2), and ear leaf plus two (EL+2).

[x] Tar spot chlorosis and necrosis symptoms were visually assessed as a percentage (0–100%) of leaf area on five plants in each plot at the EL, EL–2, and EL+2.

[w] Means followed by the same letter are not significantly different based on Fisher's Least Significant Difference test (LSD; α=0.05).

TABLE 41. ***Effect of Fungicide on Tar Spot on September 30***

TREATMENTS, RATE/ACRE, AND TIMING[z]	TAR SPOT % STROMA[y] EL−2	TAR SPOT % STROMA[y] EL	TAR SPOT % STROMA[y] EL+2	TAR SPOT % CHLOR/ NECR[x] EL−2	TAR SPOT % CHLOR/ NECR[x] EL	TAR SPOT % CHLOR/ NECR[x] EL+2
Nontreated control	39.3 a	37.0 a	30.5 a	89.3 a	76.8 a	45.5 a
Aproach 6.0 fl oz at VT	31.3 b	25.8 b	19.0 b	70.3 b	53.3 b	21.3 b
Aproach Prima 2.34 SC 6.8 fl oz at VT	21.3 c	19.5 c	14.0 c	55.8 c	28.8 c	8.1 c
Aproach 6.0 fl oz at VT/R1 fb Aproach Prima 2.34 SC 6.8 fl oz R2	16.3 d	14.8 c	5.8 e	57.5 bc	19.6 c	4.6 c
Aproach Prima 2.34 SC 6.8 fl oz at VT/R1 fb Aproach 6.0 fl oz R2	16.1 d	17.3 c	9.9 d	65.0 bc	27.0 c	6.8 c
P-value[w]	*0.0001*	*0.0001*	*0.0001*	*0.0016*	*0.0001*	*0.0001*

[z] Fungicide treatments were applied on August 8 at the tassel/silk (VT/R1) and on August 22 at the blister (R2) growth stages, and all treatments contained a nonionic surfactant (Preference) at a rate of 0.25% v/v. fb = followed by.

[y] Tar spot stroma was visually assessed as a percentage (0–100%) of leaf area on five plants in each plot at the ear leaf (EL), ear leaf minus two (EL−2), and ear leaf plus two (EL+2).

[x] Tar spot chlorosis and necrosis symptoms were visually assessed as a percentage (0–100%) of leaf area on five plants in each plot at the EL, EL−2, and EL+2.

[w] Means followed by the same letter are not significantly different based on Fisher's Least Significant Difference test (LSD; α=0.05).

TABLE 42. ***Effect of Fungicide on Canopy Greenness, Lodging, and Corn Yield***

TREATMENT, RATE/ACRE, AND TIMING[Z]	CANOPY GREEN[Y] % SEPTEMBER 21	CANOPY GREEN[Y] % SEPTEMBER 30	LODGING[X] %	HARVEST MOISTURE %	TEST WEIGHT LB/BU	YIELD[W] BU/ACRE
Nontreated control	53.8 b	32.5 c	7.5	22.7 d	52.4	186.7 c
Aproach 6.0 fl oz at VT	80.0 a	47.5 b	7.5	23.0 cd	52.4	199.3 bc
Aproach Prima 2.34 SC 6.8 fl oz at VT	86.3 a	56.3 a	0.0	23.4 bc	51.5	200.7 b
Aproach 6.0 fl oz at VT/R1 fb Aproach Prima 2.34 SC 6.8 fl oz R2	87.5 a	60.0 a	2.5	24.5 a	51.8	207.8 ab
Aproach Prima 2.34 SC 6.8 fl oz at VT/R1 fb Aproach 6.0 fl oz R2	86.3 a	62.5 a	5.0	23.8 b	52.0	213.9 a
P-value[v]	*0.0001*	*0.0001*	*0.2754*	*0.0001*	*0.4778*	*0.0071*

[z] Fungicide treatments were applied on August 8 at the tassel/silk (VT/R1) and on August 22 at the blister (R2) growth stages, and all treatments contained a nonionic surfactant (Preference) at a rate of 0.25% v/v. fb = followed by.

[y] Canopy greenness was visually assessed as a percentage (0–100%) of canopy on September 21 and September 30.

[w] Lodging was the percentage of lodged stalks when pushed from shoulder height to 45° from vertical.

[w] Yields were adjusted to 15.5% moisture at harvest on October 25.

[v] Means followed by the same letter are not significantly different based on Fisher's Least Significant Difference test (LSD; α=0.05).

FUNGICIDE COMPARISON FOR WHITE MOLD IN SOYBEAN, 2019 (SOY19-02.PPAC)

C. Rocca da Silva, S. Shim, J. D. Ravellette, and D. E. P. Telenko, Department of Botany and Plant Pathology, Purdue University West Lafayette, Indiana 47907-2054

SOYBEAN (*GLYCINE MAX* P35T75X)

White mold, *Sclerotinia sclerotiorum*
Frogeye leaf spot, *Cercospora sojina*
Cercospora leaf blight, *Cercospora kikuchii*
Septoria brown spot, *Septoria glycines*

A trial was established at the Pinney Purdue Agricultural Center (PPAC) in Porter County, IN. The experiment was a randomized complete block design with four replications. Plots were 6.7 feet wide and 30 feet long and consisted of four rows, and the two center rows were used for evaluation. The previous crop was corn. Standard practices for soybean production in Indiana were followed. Soybean variety P35T75X was planted in 20-inch row spacing at a rate of 8 seeds/ft on June 6. Inoculum of *Sclerotinia sclerotiorum*was applied on the seedbed at 1.25 g/ft at planting. The field was overhead irrigated weekly at 1 inch unless weekly rainfall was 1 inch or higher to encourage disease. All fungicide applications were applied at 15 gal/acre and 40 psi using a Lee self-propelled sprayer equipped with a 10-foot boom, fitted with six TJ-VS 8002 nozzles spaced 20 inches apart, at 3.6 mph. Fungicides were applied on July 23 at the beginning bloom (R1) and on July 30 at the full bloom (R2) growth stages. Disease ratings were assessed on September 18 at the full seed (R6) growth stage. Disease incidence was rated by visually assessing the number of symptomatic plants in each plot for white mold. Frogeye leaf spot and Septoria brown spot were rated for severity by visually assessing the percentage of symptomatic leaf area in the upper and lower canopies. The two center rows of each plot were harvested on October 23, and yields were adjusted to 13% moisture. Data were subjected to analysis of variance (SAS 9.4, 2019), and means were compared using Fisher's Least Significant Difference test (LSD; α=0.05).

In 2019, very little disease developed in plots. White mold, frogeye leaf spot, Cercospora leaf blight, and Septoria brown spot were the most prominent diseases in the trial but only reached low severity. There was no significant difference between fungicide treatments for white mold or any other diseases (Table 43). There was no significant treatment effect on canopy greenness, defoliation, and soybean yield (Table 44).

TABLE 43. ***Effect of Fungicide on Soybean Diseases***

TREATMENT AND RATE/ACRE[z]	WHITE MOLD # PLANTS/PLOT[y]	FLS[x] %	CLS[x] %	SBS[x] %
Nontreated, noninoculated	1.0	1.0	7.3	1.5
Nontreated, inoculated	0.3	1.0	5.5	1.0
Acropolis 23.0 fl oz at R2	2.5	1.0	5.5	0.8
Aproach 8.0 fl oz at R2	1.0	1.5	5.0	0.8
Endura 12.5 oz at R2	1.0	1.3	7.5	0.8
Priaxor Xemium 6.0 fl oz at R2	2.3	1.0	6.3	0.8
Stratego YLD 4.0 fl oz at R2	2.0	1.0	4.3	0.6
Proline 480 SC 4.0 fl oz at R2	0.3	1.0	5.8	0.6
Omega 0.75 pt at R2	0.5	1.5	6.8	0.8
Topsin 15.0 fl oz at R2	3.8	1.3	8.8	0.8
Miravis Neo 2.5 SE 20.8 fl oz at R1	3.5	1.0	8.0	0.6
Double Nickel 1.0 qt at R2	2.8	1.0	5.0	1.9
Double Nickel 2.0 qt at R2	2.8	2.0	8.8	0.9
P-value[w]	*0.1826*	*0.5847*	*0.4090*	*0.2681*

[z] All plots were inoculated with *S. sclerotiorum*in-furrow at planting except the nontreated, noninoculated control. Fungicide treatments were applied on July 23 at the beginning bloom (R1) and on July 30 at the full bloom (R2) growth stages. All treatments contained a nonionic surfactant at a rate of 0.25% v/v.

[y] White mold disease was assessed by counting the number of plants per plot with symptoms on September 18.

[x] Foliar disease severity was visually assessed as the percentage (0–100%) of symptomatic leaf area in the plot. FLS = frogeye leaf spot, CLB = Cercospora leaf blight, SBS = Septoria brown spot.

[w] Means followed by the same letter are not significantly different based on Fisher's Least Significant Difference test (LSD; α=0.05).

TABLE 44. ***Effect of Fungicide on Soybean Yield***

TREATMENT AND RATE/ACRE[z]	CANOPY GREEN[y] %	DEFOLIATION[x] %	HARVEST MOISTURE %	TEST WEIGHT LBS/BU	YIELD[w] BU/ACRE
Nontreated, noninoculated	92.9	2.1	12.5	55.3	80.4
Nontreated, inoculated	92.5	2.3	12.4	55.2	81.3
Acropolis 23.0 fl oz at R2	90.0	2.5	12.5	55.1	79.5
Aproach 8.0 fl oz at R2	92.5	2.5	12.7	55.3	82.9
Endura 12.5 oz at R2	92.5	1.8	12.6	55.4	80.6
Priaxor Xemium 6.0 fl oz at R2	92.5	2.5	12.7	55.5	81.1
Stratego YLD 4.0 fl oz at R2	96.3	1.5	12.6	55.2	82.2
Proline 480 SC 4.0 fl oz at R2	92.5	2.0	12.6	55.2	78.0
Omega 0.75 pt at R2	92.5	2.8	12.7	55.4	80.6
Topsin 15.0 fl oz at R2	87.5	2.5	12.6	55.4	81.9
Miravis Neo 2.5 SE 20.8 fl oz at R1	91.3	2.8	13.0	55.3	79.3
Double Nickel 1.0 qt at R2	87.5	3.8	12.8	55.1	79.8
Double Nickel 2.0 qt at R2	87.5	2.5	12.6	55.2	77.7
P-value[v]	*0.5527*	*0.8474*	*0.5915*	*0.3664*	*0.8352*

[z] All plots were inoculated with *S. sclerotiorum* in-furrow at planting except the nontreated, noninoculated control. Fungicide treatments were applied on July 23 at the beginning bloom (R1) and July 30 at the full bloom (R2) growth stages. All treatments contained a nonionic surfactant at a rate of 0.25% v/v.

[y] Canopy green was visually assessed as the percentage (0–100%) in the plot on September 18.

[x] Defoliation was visually assessed as percentage of leaf loss in the plot.

[w] Yields were adjusted to 13% moisture at harvest on October 23.

[v] Means followed by the same letter are not significantly different based on Fisher's Least Significant Difference (LSD; α=0.05).

EVALUATION OF FUNGICIDE EVALUATION FOR FOLIAR DISEASES IN SOYBEAN IN NORTHWESTERN INDIANA, 2019 (SOY19-14.PPAC)

C. Rocca da Silva, S. Shim, J. D. Ravellette, and D. E. P. Telenko, Department of Botany and Plant Pathology, Purdue University West Lafayette, Indiana 47907-2054

SOYBEAN (*GLYCINE MAX* P34A13X)

White mold, *Sclerotinia sclerotiorum*
Frogeye leaf spot, *Cercospora sojina*
Cercospora leaf blight, *Cercospora kikuchii*
Septoria brown spot, *Septoria glycines*

A trial was established at the Pinney Purdue Agricultural Center (PPAC) in Porter County, IN. The experiment was a randomized complete block design with four replications. Plots were 10 feet wide and 30 feet long and consisted of four rows, and the two center rows were used for evaluation. The previous crop was corn. Standard practices for soybean production in Indiana were followed. Soybean variety P34A13X was planted in 30-inch row spacing at a rate of 140,000 seeds/acre on June 8. The field was overhead irrigated weekly at 1 inch unless weekly rainfall was 1 inch or higher to encourage disease. All fungicide applications were applied at 15 gal/acre and 40 psi using a Lee self-propelled sprayer equipped with a 10-foot boom, fitted with six TJ-VS 8002 nozzles spaced 20 inches apart, at 3.6 mph. Fungicides were applied on August 10 at the beginning pod (R3) growth stage. Disease ratings were assessed on September 18 at the full pod (R6) growth stage. Disease severity was rated by visually assessing the number of symptomatic plants in each plot for white mold. Frogeye leaf spot, Cercospora leaf blight, and Septoria brown spot were rated for disease severity by visually assessing the percentage of symptomatic leaf area in the upper and lower canopies on September 18. The two center rows of each plot were harvested on October 24, and yields were adjusted to 13% moisture. Data were subjected to analysis of variance (SAS 9.4, 2019), and means were compared using Fisher's Least Significant Difference test (LSD; α=0.05).

In 2019, very little disease developed in plots. White mold, frogeye leaf spot, Septoria brown spot, and Cercospora leaf blight were the most prominent diseases in the trial but only reached low severity. There was no significant difference between fungicide treatments and the nontreated control for all disease ratings on September 18 (Table 45). There was no significant effect of treatment on canopy greenness, defoliation or soybean yield (Table 46).

TABLE 45. ***Effect of Fungicide on Soybean Diseases***

TREATMENT AND RATE/ACRE[z]	WHITE MOLD # PLANTS/PLOT[y]	FLS[x] %	CLS[x] %	SBS[x] %
Nontreated control	5.0	1.3	0.8	1.9
Miravis Top 1.67 SC 13.7 fl oz	4.0	1.0	3.0	0.3
Miravis Neo 2.5 SE 13.7 fl oz	2.5	1.0	1.3	0.4
Delaro Complete 458 SC 8.0 fl oz	2.8	1.3	1.3	0.3
Delaro 325 SC 8.0 fl oz	4.8	1.0	0.0	0.2
Lucento 4.17 SC 5.0 fl oz	10.0	1.0	0.5	0.4
Aproach Prima 2.34 SC 6.8 fl oz	5.0	1.0	1.5	0.5
Priaxor Xemium 4.0 fl oz	18.5	1.3	1.8	0.3
Headline AMP 1.68 SC 10.0 fl oz	21.3	2.0	1.5	0.5
Veltyma 3.34 SC 7.0 fl oz	3.3	1.0	3.0	0.4
P-value[w]	*0.3409*	*0.5403*	*0.2991*	*0.1155*

[z] Fungicide treatments were applied on August 10 at the beginning pod (R3) growth stage. All treatments contained a nonionic surfactant at a rate of 0.25% v/v.

[y] White mold disease was assessed by counting the number of plants per plot with symptoms on September 18.

[x] Foliar disease severity was visually assessed as the percentage (0–100%) of symptomatic leaf area in the plot on September 18. FLS = frogeye leaf spot, CLB = Cercospora leaf blight, SBS = Septoria brown spot.

[w] Means followed by the same letter are not significantly different based on Fisher's Least Significant Difference test (LSD; α=0.05).

TABLE 46. ***Effect of Fungicide on Soybean Yield***

TREATMENT AND RATE/ACRE[z]	CANOPY GREEN[y] %	DEFOLIATION[x] %	HARVEST MOISTURE %	TEST WEIGHT LBS/BU	YIELD[w] BU/ACRE
Nontreated control	92.5	1.5	13.9	55.6	63.1
Miravis Top 1.67 SC 13.7 fl oz	95.0	1.0	13.8	55.5	67.4
Miravis Neo 2.5 SE 13.7 fl oz	95.0	1.0	13.9	55.3	64.8
Delaro Complete 458 SC 8.0 fl oz	93.8	1.3	13.8	55.6	63.7
Delaro 325 SC 8.0 fl oz	95.0	1.0	13.9	55.5	63.2
Lucento 4.17 SC 5.0 fl oz	95.0	1.3	13.7	55.6	63.5
Aproach Prima 2.34 SC 6.8 fl oz	95.0	1.3	13.7	55.4	64.6
Priaxor Xemium 4.0 fl oz	88.8	2.0	13.9	55.6	62.6
Headline AMP 1.68 SC 10.0 fl oz	90.0	2.5	14.1	55.4	63.6
Veltyma 3.34 SC 7.0 fl oz	93.8	1.0	14.7	56.2	64.2
P-value[v]	*0.4699*	*0.2256*	*0.5023*	*0.5956*	*0.5187*

[z] Fungicide treatments were applied on August 10 at the beginning pod (R3) growth stage. All treatments contained a nonionic surfactant at a rate of 0.25% v/v.

[y] Canopy green was visually assessed as the percentage (0–100%) in the plot on September 18.

[x] Defoliation was visually assessed as a percentage of leaf loss in the plot.

[w] Yields were adjusted to 13% moisture at harvest on October 24.

[v] Means followed by the same letter are not significantly different based on Fisher's Least Significant Difference (LSD; α=0.05).

COMPARISON OF FUNGICIDES FOR WHITE MOLD IN SOYBEAN IN NORTHWESTERN INDIANA, 2019 (SOY19-15.PPAC)

S. Shim, J. D. Ravellette, and D. E. P. Telenko, Department of Botany and Plant Pathology, Purdue University West Lafayette, Indiana 47907-2054

SOYBEAN (*GLYCINE MAX* P35T75X)

White mold, *Sclerotinia sclerotiorum*
Frogeye leaf spot, *Cercospora sojina*
Cercospora leaf blight, *Cercospora kikuchii*
Septoria brown spot, *Septoria glycines*

A trial was established at the Pinney Purdue Agricultural Center (PPAC) in Porter County, IN. The experiment was a randomized complete block design with four replications. Plots were 6.7 feet wide and 30 feet long and consisted of four rows, and the two center rows were used for evaluation. The previous crop was corn. Standard practices for soybean production in Indiana were followed. Soybean variety P35T75X was planted in 15-inch row spacing at a rate of 8 seeds/ft on June 6. Inoculum of *S. sclerotiorum*was applied on the seedbed at 1.25 g/ft at planting. The field was overhead irrigated weekly at 1 inch unless weekly rainfall was 1 inch or higher to encourage disease. All fungicide applications were applied at 15 gal/acre and 40 psi using a Lee self-propelled sprayer equipped with a 10-foot boom, fitted with six TJ-VS 8002 nozzles spaced 20 inches apart, at 3.6 mph. Fungicides were applied on July 23 at the beginning bloom (R1) and on August 10 at the beginning pod (R3) growth stages. Disease ratings were assessed on September 18 at the full seed (R6) growth stage. Disease severity was rated by visually assessing the number of symptomatic plants in each plot for white mold. Frogeye leaf spot, Cercospora leaf blight, and Septoria brown spot were rated for disease severity by visually assessing the percentage of symptomatic leaf area in the upper and lower canopies. The two center rows of each plot were harvested on October 23, and yields were adjusted to 13% moisture. Data were subjected to analysis of variance (SAS 9.4, 2019), and means were compared using Fisher's Least Significant Difference test (LSD; α=0.05).

In 2019, very little disease developed in plots. White mold, frogeye leaf spot, sudden death syndrome, Septoria brown spot, and Cercospora leaf blight were the most prominent diseases but only reached low severity. There was no significant difference between fungicide treatments and the nontreated control for all diseases rated on September 18 (Table 47). There was no significant effect of treatment on canopy greenness, defoliation, or soybean yield (Table 48).

TABLE 47. ***Effect of Fungicide on Soybean Diseases***

TREATMENT AND RATE/ACRE[z]	WHITE MOLD % DI[Y]	FLS[X] %	CLS[X] %	SBS[X] %
Nontreated control	3.3	1.0	8.8	0.8
Proline 480 SC 3.0 fl oz at R1	2.9	1.0	8.8	0.5
Delaro 325 SC 8.0 fl oz at R1	0.8	1.0	6.3	0.5
Delaro Complete 458 SC 8.0 fl oz at R1	1.3	1.0	10.0	0.6
Delaro 325 SC 8.0 fl oz at R1 fb R3	1.3	1.0	7.0	0.5
Delaro Complete 458 SC 8.0 fl oz at R1 fb R3	3.8	1.0	7.5	0.5
P-value[w]	*0.8087*	-	*0.7771*	*0.2161*

[z] All plots were inoculated with *S. sclerotiorum* in-furrow at planting. Fungicides were applied on July 23 at the beginning bloom (R1) and on August 10 at the beginning pod (R3) growth stages. All treatments contained a nonionic surfactant at a rate of 0.25% v/v. fb = followed by.

[y] White mold disease incidence was assessed by counting the number of plants per plot with symptoms on September 18.

[x] Foliar disease severity was visually assessed as a percentage (0–100%) of symptomatic leaf area in the plot. FLS = frogeye leaf spot, CLB = Cercospora leaf blight, SBS = Septoria brown spot.

[w] Means followed by the same letter are not significantly different based on Fisher's Least Significant Difference test (LSD; α=0.05).

TABLE 48. ***Effect of Fungicide on Soybean Yield***

TREATMENT AND RATE/ACRE[z]	CANOPY GREEN[y] %	DEFOLIATION[x] %	HARVEST MOISTURE %	TEST WEIGHT LBS/BU	YIELD[w] BU/ACRE
Nontreated control	87.5	3.0	13.4	55.7	109.6
Proline 480 SC 3.0 fl oz at R1	90.0	2.0	13.4	55.6	106.9
Delaro 325 SC 8.0 fl oz at R1	90.0	2.3	14.3	56.2	112.0
Delaro Complete 458 SC 8.0 fl oz at R1	91.3	3.0	14.4	54.9	110.2
Delaro 325 SC 8.0 fl oz at R1 fb R3	88.8	2.8	13.5	55.5	113.8
Delaro Complete 458 SC 8.0 fl oz at R1 fb R3	90.0	2.8	13.4	56.3	113.4
P-value[v]	*0.8319*	*0.9072*	*0.3439*	*0.2392*	*0.3670*

[z] All plots were inoculated with *S. sclerotiorum* in-furrow at planting. Fungicides were applied on July 23 at the beginning bloom (R1) and on August 10 at the beginning pod (R3) growth stages. All treatments contained a nonionic surfactant at a rate of 0.25% v/v. fb = followed by.

[y] Canopy green was visually assessed as a percentage (0–100%) in the plot on September 18.

[x] Defoliation was visually assessed as a percentage of leaf loss in the plot.

[w] Yields were adjusted to 13% moisture at harvest on October 23.

[v] Means followed by the same letter are not significantly different based on Fisher's Least Significant Difference (LSD; α=0.05).

FUNGICIDE EVALUATION FOR WHITE MOLD IN SOYBEAN IN NORTHWESTERN INDIANA, 2019 (SOY19-22.PPAC)

C. Rocco da Silva, S. Shim, J. D. Ravellette, and D. E. P. Telenko, Department of Botany and Plant Pathology, Purdue University West Lafayette, Indiana 47907-2054

SOYBEAN (*GLYCINE MAX* P35T75X)

White mold, *Sclerotinia sclerotiorum*
Frogeye leaf spot, *Cercospora sojina*
Septoria brown spot, *Septoria glycines*

A trial was established at the Pinney Purdue Agricultural Center (PPAC) in Porter County, IN. The experiment was a randomized complete block design with four replications. Plots were 6.7 feet wide and 30 feet long and consisted of four rows, and the two center rows were used for evaluation. The previous crop was corn. Standard practices for soybean production in Indiana were followed. Soybean variety P35T75X was planted in 20-inch row spacing at a rate of 8 seeds/ft on June 6. Inoculum of *S. sclerotiorum*was applied on the seedbed at 1.25 g/ft at planting. The field was overhead irrigated weekly at 1 inch unless weekly rainfall was 1 inch or higher to encourage disease. All fungicide applications were applied at 15 gal/acre and 40 psi using a Lee self-propelled sprayer equipped with a 10-foot boom, fitted with six TJ-VS 8002 nozzles spaced 20 inches apart, at 3.6 mph. Fungicides were applied on July 30 at the beginning bloom (R1) growth stage. Disease ratings were assessed on September 18 at the full seed (R6) growth stage. Disease severity was rated by visually assessing the number of symptomatic plants in each plot for white mold. Frogeye leaf spot and Septoria brown spot were rated for severity by visually assessing the percentage of symptomatic leaf area in the upper and lower canopies. The two center rows of each plot were harvested on October 23, and yields were adjusted to 13% moisture. Data were subjected to analysis of variance (SAS 9.4, 2014), and means were compared using Fisher's Least Significant Difference test (LSD; α=0.05).

In 2019, very little disease developed in plots. White mold, frogeye leaf spot, and Septoria brown spot were the most prominent diseases in the trial but only reached low severity. There was no significant difference between fungicide treatments and the nontreated control for frogeye leaf spot, Cercospora leaf blight, and Septoria brown spot on September 18 (Table 49). All fungicide programs reduced the number of plants with white mold over the nontreated control on September 18 except Domark plus Badge. There was no significant treatment effect on defoliation, harvest moisture, test weight, and soybean yield (Table 50).

TABLE 49. ***Effect of Fungicide on Soybean Diseases***

TREATMENT AND RATE/ACRE[z]	WHITE MOLD % DI[y]	FLS[x] %	CLS[x] %	SBS[x] %
Nontreated control	2.8 a	1.5	6.5	1.5
Domark 5.0 fl oz	0.5 cd	1.0	5.0	1.0
Domark 5.0 fl oz + Badge 1.0 qt	2.3 ab	1.0	5.0	1.0
Aproach 10.0 fl oz	0.3 cd	1.3	5.5	1.0
Aproach 10.0 fl oz + Badge 1.0 qt	0.0 d	1.3	5.0	2.3
Endura 8.0 oz	0.5 cd	1.0	5.0	1.0
Endura 8.0 oz + Badge 1.0 qt	0.3 cd	1.0	5.5	1.0
Topsin 15.0 fl oz	0.0 d	1.0	4.3	1.0
Topsin 15.0 fl oz + Badge 1.0 qt	0.5 cd	1.3	4.5	2.0
Badge 1.0 qt	1.8 abc	1.0	4.0	1.3
GWN 10473 6.0 fl oz	0.0 d	1.0	7.0	0.9
GWN 10473 6.0 fl oz + Badge 1.0 qt	0.8 bcd	1.0	3.5	1.0
P-value[w]	*0.0068*	*0.3526*	*0.7807*	*0.3664*

[z] All plots were inoculated with *S. sclerotiorum* in-furrow at planting. Fungicides were applied on July 30 at the beginning bloom (R1).

[y] White mold disease incidence was assessed by counting the number of plants per plot with symptoms on September 18.

[x] Foliar disease severity was visually assessed as a percentage (0–100%) of symptomatic leaf area in the plot. FLS = frogeye leaf spot, CLB = Cercospora leaf blight, SBS = Septoria brown spot.

[w] Means followed by the same letter are not significantly different based on Fisher's Least Significant Difference test (LSD; α=0.05).

TABLE 50. ***Effect of Fungicide on Soybean Yield***

TREATMENT AND RATE/ACRE[z]	DEFOLIATION[y] %	HARVEST MOISTURE %	TEST WEIGHT LBS/BU	YIELD[x] BU/ACRE
Nontreated control	4.0	12.3	55.5	80.0
Domark 5.0 fl oz	2.0	12.5	55.1	76.2
Domark 5.0 fl oz + Badge 1.0 qt	1.0	12.3	55.2	78.0
Aproach 10.0 fl oz	1.0	12.4	55.3	78.2
Aproach 10.0 fl oz + Badge 1.0 qt	3.3	12.5	55.3	75.3
Endura 8.0 oz	3.3	12.5	55.5	76.4
Endura 8.0 oz + Badge 1.0 qt	2.5	12.3	55.3	83.2
Topsin 15.0 fl oz	2.5	12.5	55.3	75.4
Topsin 15.0 fl oz + Badge 1.0 qt	1.3	12.3	55.2	78.9
Badge 1.0 qt	2.0	12.4	55.3	78.3
GWN 10473 6.0 fl oz	2.8	12.6	55.1	77.4
GWN 10473 6.0 fl oz + Badge 1.0 qt	2.0	12.6	55.2	75.0
P-value[w]	*0.1443*	*0.7999*	*0.6037*	*0.8026*

[z] All plots were inoculated with *S. sclerotiorum* in-furrow at planting. Fungicides were applied on July 30 at the beginning bloom (R1).

[y] Defoliation was visually assessed as a percentage of leaf loss in plot.

[x] Yields were adjusted to 13% moisture at harvest on October 23.

[w] Means followed by the same letter are not significantly different based on Fisher's Least Significant Difference (LSD; α=0.05).

SOUTHWEST PURDUE AGRICULTURAL CENTER (SWPAC)

COMPARISON OF FUNGICIDES APPLIED AT SILK (R1) AND MILK (R3) GROWTH STAGES FOR FOLIAR DISEASES IN CORN IN SOUTHWESTERN INDIANA, 2019 (COR19-18.SWPAC)

C. Rocca da Silva, S. Shim, J. D. Ravellette, and D. E. P. Telenko, Department of Botany and Plant Pathology, Purdue University West Lafayette, Indiana 47907-2054

CORN (*ZEA MAYS* P9998AM)

Southern rust, *Puccinia polysora*
Northern corn leaf blight, *Exserohilum turcicum*
Gray leaf spot, *Cercospora zeae-maydis*

A trial was established at the Southwest Purdue Agricultural Center (SWPAC) in Knox County, IN. The experiment was a randomized complete block design with four replications. Plots were 10 feet wide and 30 feet long and consisted of four rows, and the two center rows were used for evaluation. The previous crop was corn. Standard practices for nonirrigated grain corn production in Indiana were followed. Corn hybrid P9998AM was planted in 30-inch row spacing at a rate of 27,000 seeds/acre on May 28. All fungicide applications were applied at 15 gal/acre and 40 psi using a Lee self-propelled sprayer equipped with a 10-foot boom, fitted with six TJ-VS 8002 nozzles spaced 20 inches apart, at 3.6 mph. Fungicides were applied on July 24 at the early silk (R1) growth stage and on August 14 at the milk (R3) growth stage. Disease ratings were assessed on August 27 at the dough (R4) growth stage and on September 12 at the dent (R5) growth stage. Disease severity was rated by visually assessing the percentage of symptomatic leaf area of the ear leaf on five plants in each plot. Values for each plot were averaged before analysis. The two center rows of each plot were harvested on October 9, and yields were adjusted to 15.5% moisture. Data were subjected to analysis of variance (SAS 9.4, 2019), and means were compared using Fisher's Least Significant Difference test (LSD; α=0.05).

In 2019, weather conditions were favorable for disease. Southern rust, gray leaf spot, and northern corn leaf blight were the most prominent diseases in the trial and reached moderate severity. Physoderma brown spot was also noted in some plots but was not rated. Lucento, Trivapro, Miravis Neo, Veltyma, Delaro 325 SC, Quilt Xcel, and Revytek applied at the early R1 growth stage significantly reduced southern rust on August 27 compared to the nontreated control (Table 51). There was no significant difference between fungicide treatments and the nontreated control for northern corn leaf blight and gray leaf spot on August 27. All fungicides at the R3 application timings significantly reduced southern rust compared to the nontreated control on September 12. In addition, all fungicide treatments and timings reduced gray leaf spot on September 12. By September 12, only Lucento and Revytek applied at the early R1 growth stage had significantly less northern corn leaf blight than the nontreated control. On September 12, all fungicides applied at R3 reduced northern corn leaf blight over the nontreated control except Miravis Neo. There was no significant difference between treatments for percentage of canopy greenness, lodging, and ear rot (Table 52). Test weight was significantly lower than the nontreated control for Veltyma, Quilt Xcel, Headline AMP, and Revytek applied at the early R1 growth stage and Veltyma, Quilt Xcel, and Revytek applied at R3, but no significant differences were detected between fungicide treatments and the nontreated control for yield.

TABLE 51. ***Effect of Fungicide on Foliar Disease Severity***

TREATMENT, RATE/ACRE, AND TIMING[Z]	SR %[Y] AUGUST 27	NCLB %[Y] AUGUST 27	GLS %[Y] AUGUST 27	SR %[Y] SEPTEMBER 12	NCLB %[Y] SEPTEMBER 12	GLS %[Y] SEPTEMBER 12
Nontreated control	0.7 a	0.0	2.5	12.0 a	7.3 a	4.0 a
Lucento 4.17 SC 5.0 fl oz at VT/R1	0.2 bc	0.1	1.3	0.7 b	1.3 de	1.0 cd
Trivapro 2.21 SE 13.7 fl oz at VT/R1	0.1 c	0.2	1.1	0.3 b	5.6 abc	1.3 cd
Miravis Neo 2.5 SE 13.7 fl oz at VT/R1	0.1 c	0.0	1.2	1.2 b	5.0 a-d	0.7 d
Veltyma 3.34 SC 7.0 fl oz at VT/R1	0.1 c	0.2	1.2	0.8 b	0.8 e	0.8 d
Delaro 325 SC 8.0 fl oz at VT/R1	0.1 c	0.1	1.5	2.5 b	6.7 ab	1.2 cd
Quilt Xcel 2.2 SE 10.5 fl oz at VT/R1	0.0 c	0.5	1.2	1.4 b	5.5 a-d	1.2 cd
Headline AMP 1.68 SC 10.0 fl oz at VT/R1	0.4 abc	0.1	1.3	0.9 b	5.9 abc	1.3 cd
Revytek 3.33 LC 8.0 fl oz at VT/R1	0.1 c	0.2	1.3	3.7 b	0.4 e	0.8 d
Lucento 4.17 SC 5.0 fl oz at R3	0.6 ab	0.7	1.7	0.7 b	0.6 e	1.8 bcd
Trivapro 2.21 SE 13.7 fl oz at R3	0.3 abc	0.0	2.6	0.4 b	2.3 cde	1.8 bcd
Miravis Neo 2.5 SE 13.7 fl oz at R3	0.4 abc	0.3	1.7	0.9 b	3.6 a-e	2.2 bc
Veltyma 3.34 SC 7.0 fl oz at R3	0.7 a	0.5	1.7	0.5 b	0.5 e	0.9 cd
Delaro 325 SC 8.0 fl oz at R3	0.3 abc	0.3	2.0	1.3 b	3.0 b-e	1.7 bcd
Quilt Xcel 2.2 SE 10.5 fl oz at R3	0.3 abc	0.2	1.7	0.5 b	2.3 cde	2.6 b
Headline AMP 1.68 SC 10.0 fl oz at R3	0.3 abc	0.6	1.7	1.7 b	0.0 e	1.7 bcd
Revytek 3.33 LC 8.0 fl oz at R3	0.4 abc	0.9	2.3	2.9 b	3.0 b-e	1.4 bcd
P-value[x]	*0.0130*	*0.3483*	*0.2247*	*0.0001*	*0.0041*	*0.0008*

[z] Fungicide treatments were applied on July 24 at the early R1 (silk) growth stage and on August 14 at the milk (R3) growth stage, and all treatments contained a nonionic surfactant at a rate of 0.25% v/v.

[y] Disease severity was visually assessed as a percentage (0–100%) of symptomatic leaf area on ear leaves. Five leaves were assessed per plot and averaged. SR = southern rust, NCLB = northern corn leaf blight, GLS = gray leaf spot.

[x] Means followed by the same letter are not significantly different based on Fisher's Least Significant Difference test (LSD; α=0.05).

TABLE 52. ***Effect of Fungicide on Canopy Greenness, Lodging, and Corn Yield***

TREATMENT, RATE/ACRE, AND TIMING[z]	CANOPY GREEN[y] %	LODGING[x] %	EAR ROT[w] %	HARVEST MOISTURE %	TEST WEIGHT LB/BU	YIELD[v] BU/ACRE
Nontreated control	60.0	2.5	4.5	14.0	58.2 a	154.1
Lucento 4.17 SC 5.0 fl oz at VT/R1	56.3	0.0	4.9	14.5	58.1 a	149.0
Trivapro 2.21 SE 13.7 fl oz at VT/R1	48.8	0.0	3.7	14.0	57.7 a-c	154.6
Miravis Neo 2.5 SE 13.7 fl oz at VT/R1	57.5	0.0	3.2	14.5	57.6 abc	157.5
Veltyma 3.34 SC 7.0 fl oz at VT/R1	51.3	0.0	5.0	14.0	57.4 b-d	163.9
Delaro 325 SC 8.0 fl oz at VT/R1	63.8	5.0	4.4	14.2	57.7 abc	167.7
Quilt Xcel 2.2 SE 10.5 fl oz at VT/R1	51.3	0.0	3.6	14.4	56.9 e	155.6
Headline AMP 1.68 SC 10.0 fl oz at VT/R1	60.0	0.0	4.0	14.5	57.4 b-e	151.9
Revytek 3.33 LC 8.0 fl oz at VT/R1	57.5	0.0	3.7	14.7	57.3 b-e	152.0
Lucento 4.17 SC 5.0 fl oz at R3	53.8	0.0	4.0	14.2	57.6 a-d	168.1
Trivapro 2.21 SE 13.7 fl oz at R3	48.8	0.0	3.2	14.4	57.5 a-e	165.7
Miravis Neo 2.5 SE 13.7 fl oz at R3	55.0	2.5	3.5	14.5	57.8 abc	161.1
Veltyma 3.34 SC 7.0 fl oz at R3	55.0	2.5	4.1	14.4	57.0 de	155.1
Delaro 325 SC 8.0 fl oz at R3	50.0	0.0	5.4	14.5	57.8 ab	155.7
Quilt Xcel 2.2 SE 10.5 fl oz at R3	58.8	0.0	4.7	14.5	57.2 cde	163.4
Headline AMP 1.68 SC 10.0 fl oz at R3	62.5	2.5	4.5	14.3	57.6 a-d	162.8
Revytek 3.33 LC 8.0 fl oz at R3	52.5	0.0	4.4	14.6	57.4 b-e	161.0
P-value[u]	*0.5601*	*0.6716*	*0.6780*	*0.8951*	0.0110	*0.6268*

[z] Fungicide treatments were applied on July 24 at the early R1 (silk) growth stage and on August 14 at the milk (R3) growth stage, and all treatments contained a nonionic surfactant at a rate of 0.25% v/v.

[y] Canopy greenness was visually assessed as a percentage (0–100%) on September 12.

[x] Lodging is the percentage of lodged stalks when pushed from shoulder height to 45° from vertical on September 12.

[w] Ear rot was visually assessed as a percentage (0–100%) from 10 ears per plot. A mix of ear rot pathogens was identified and included *Fusarium*spp., *Gibberella, Diplodia*, and *Trichoderma*associated with significant insect feeding (data not presented).

[v] Yields were adjusted to 15.5% moisture at harvest on October 9.

[u] Means followed by the same letter are not significantly different based on Fisher's Least Significant Difference test (LSD; α=0.05).

EVALUATION OF FUNGICIDES FOR FOLIAR DISEASES IN CORN IN SOUTHWESTERN INDIANA, 2019 (COR19-20.SWPAC)

S. Shim, J. D. Ravellette, and D. E. P. Telenko, Department of Botany and Plant Pathology, Purdue University
West Lafayette, Indiana 47907-2054

CORN (*ZEA MAYS* P9998AM)

Southern rust, *Puccinia polysora*
Northern corn leaf blight, *Exserohilum turcicum*
Gray leaf spot, *Cercospora zeae-maydis*

A trial was established at the Southwest Purdue Agricultural Center (SWPAC) in Knox County, IN. The experiment was a randomized complete block design with four replications. Plots were 10 feet wide and 30 feet long and consisted of four rows, and the two center rows were used for evaluation. The previous crop was corn. Standard practices for nonirrigated grain corn production in Indiana were followed. Corn hybrid P9998AM was planted in 30-inch row spacing at a rate of 27,000 seeds/acre on May 28. All fungicide applications were applied at 15 gal/acre and 40 psi using a Lee self-propelled sprayer equipped with a 10-foot boom, fitted with six TJ-VS 8002 nozzles spaced 20 inches apart, at 3.6 mph. Fungicides were applied on July 24 at the early R1 (silk) growth stage. Disease ratings were assessed on August 27 at the dough (R4) growth stage and on September 12 at the dent (R5) growth stage. Disease severity was rated by visually assessing the percentage of symptomatic leaf area of the ear leaf on five plants in each plot. Values for each plot were averaged before analysis. The two center rows of each plot were harvested on October 9, and yields were adjusted to 15.5% moisture. Data were subjected to analysis of variance (SAS 9.4, 2019), and means were compared using Fisher's Least Significant Difference test (LSD; α=0.05).

In 2019, weather conditions were favorable for disease. Southern rust, gray leaf spot, and northern corn leaf blight were the most prominent diseases in the trial and reached moderate severity. Physoderma brown spot was also noted in some plots but was not rated. All fungicide treatments reduced gray leaf spot compared to the nontreated control on August 27 and September 12 (Table 53). Brixen, Delaro Complete, and Headline AMP treatments were not significantly different from the nontreated control for southern rust on August 27, but on September 12 all fungicides treatments had significantly less southern rust compared to the nontreated control. There was no difference between treatments for northern corn leaf blight on August 27, but on September 12 all fungicides significantly reduced northern corn leaf blight over the nontreated control except Delaro Complete. No significant differences were detected between fungicide treatments and the nontreated control for canopy greenness, lodging, and corn yield (Table 54).

TABLE 53. ***Effect of Fungicide on Foliar Disease Severity***

TREATMENT, RATE/ACRE, AND TIMING[z]	SR %[Y] AUGUST 27	NCLB %[Y] AUGUST 27	GLS %[Y] AUGUST 27	SR %[Y] SEPTEMBER 12	NCLB %[Y] SEPTEMBER 12	GLS %[Y] SEPTEMBER 12
Nontreated control	4.3 a	1.1 a	1.4	4.0 a	6.3 a	9.3 a
Miravis Neo 2.5 SE 13.7 fl oz	0.7 b	0.4 bc	0.1	0.7 b	1.6 b	3.4 bc
Trivapro 2.21 SE 13.7 fl oz	1.0 b	0.1 c	0.8	0.5 b	0.5 b	1.9 c
Brixen 13.7 fl oz	1.1 b	0.7 ab	0.0	0.3 b	0.9 b	1.0 c
Fortix 5.0 fl oz	0.8 b	0.5 bc	0.0	0.6 b	0.7 b	1.2 c
Delaro Complete 458 SC 8.0 fl oz	0.8 b	0.8 ab	0.4	0.6 b	1.0 b	6.8 ab
Veltyma 3.34 SC 7.0 fl oz	1.0 b	0.3 bc	0.8	0.5 b	0.6 b	0.1 c
Quilt Xcel 2.2 SE 10.5 fl oz	1.0 b	0.4 bc	0.0	0.3 b	0.7 b	2.3 c
Headline AMP 1.68 SC 10.5 fl oz	0.8 b	0.7 ab	0.6	0.7 b	0.9 b	3.2 bc
P-value[x]	*0.0001*	*0.0162*	*0.1045*	*0.0001*	*0.0051*	*0.0007*

[z] Fungicide treatments were applied on July 24 at the tassel/silk (VT/R1) growth stage, and all treatments contained a nonionic surfactant at a rate of 0.25% v/v.

[y] Disease severity was visually assessed as a percentage (0–100%) of symptomatic leaf area on ear leaves. Five leaves were assessed per plot and averaged. SR = southern rust, NCLB = northern corn leaf blight, GLS = gray leaf spot.

[x] Means followed by the same letter are not significantly different based on Fisher's Least Significant Difference test (LSD; α=0.05).

TABLE 54. ***Effect of Fungicide on Canopy Greenness, Lodging, and Corn Yield***

TREATMENT, RATE/ACRE, AND TIMING[z]	CANOPY GREEN[y] %	LODGING[x] %	HARVEST MOISTURE %	TEST WEIGHT LB/BU	YIELD[w] BU/ACRE
Nontreated control	33.8	0.3	14.3	57.2	145.1
Miravis Neo 2.5 SE 13.7 fl oz	52.5	0.3	14.7	57.4	148.1
Trivapro 2.21 SE 13.7 fl oz	40.0	0.0	14.2	57.6	146.4
Brixen 13.7 fl oz	43.8	0.3	14.3	57.5	158.2
Fortix 5.0 fl oz	40.0	0.5	14.4	57.3	137.7
Delaro Complete 458 SC 8.0 fl oz	51.3	0.5	14.5	57.2	156.4
Veltyma 3.34 SC 7.0 fl oz	45.0	0.0	15.2	57.0	159.3
Quilt Xcel 2.2 SE 10.5 fl oz	37.5	0.0	14.6	57.0	150.6
Headline AMP 1.68 SC 10.5 fl oz	48.8	0.5	14.2	57.4	149.9
P-value[v]	*0.0615*	*0.4613*	*0.0537*	*0.0438*	*0.6725*

[z] Fungicide treatments were applied on July 24 at the tassel/silk (VT/R1) growth stage, and all treatments contained a nonionic surfactant at a rate of 0.25% v/v.

[y] Canopy greenness was visually assessed as a percentage (0–100%) of canopy on September 12.

[x] Lodging is the percentage of lodged stalks when pushed from shoulder height to 45° from vertical on September 12.

[w] Yields were adjusted to 15.5% moisture at harvest on October 9.

[v] Means followed by the same letter are not significantly different based on Fisher's Least Significant Difference test (LSD; α=0.05).

EVALUATION OF FUNGICIDES FOR FOLIAR DISEASES ON SOYBEAN IN SOUTHWESTERN INDIANA, 2019 (SOY19-18.SWPAC)

C. Haupt, S. Shim, J. D. Ravellette, and D. E. P. Telenko, Department of Botany and Plant Pathology, Purdue University West Lafayette, Indiana 47907-2054

SOYBEAN (*GLYCINE MAX* P35T75X)

Frogeye leaf spot, *Cercospora sojina*
Septoria Brown spot, *Septoria glycines*
Cercospora leaf blight, *Cercospora kikuchii*

A trial was established at the Southwest Purdue Agricultural Center (SWPAC) in Knox County, IN. The experiment was a randomized complete block design with four replications. Plots were 10 feet wide and 30 feet long and consisted of four rows, and the two center rows were used for evaluation. The previous crop was soybean. Standard practices for nonirrigated soybean production in Indiana were followed. Soybean variety P35T75X was planted in 30-inch row spacing at a rate of 150,000 seeds/acre on June 3. All fungicide applications were applied at 15 gal/acre and 40 psi using a Lee self-propelled sprayer equipped with a 10-foot boom, fitted with six TJ-VS 8002 nozzles spaced 20 inches apart, at 3.6 mph. Fungicides were applied on August 1 at the beginning pod (R3) growth stage. Disease ratings were assessed on August 27 at the beginning seed (R5) growth stage and on September 12 at the full seed (R6) growth stage. Frogeye leaf spot, Cercospora leaf blight, and Septoria brown spot were rated by visually assessing the percentage of symptomatic leaf area in the upper and lower canopies. The two center rows were harvested on October 8, and yields were adjusted to 13% moisture. Data were subjected to analysis of variance (SAS 9.4, 2019), and means were compared using Fisher's Least Significant Difference test (LSD; α=0.05).

In 2019, frogeye leaf spot, Septoria brown spot, and Cercospora leaf blight were the most prominent diseases in the trial and reached low severity. All fungicide treatments reduced frogeye leaf spot and Septoria brown spot in both the upper and lower canopies compared to the nontreated control on August 27 and September 12 (Tables 55 and 56). Veltyma, Headline AMP, Miravis Neo, Miravis Top, and Delaro 325 SC reduced Cercospora leaf blight severity compared to the nontreated control on September 12 (Table 55). All fungicides reduced defoliation as compared to the nontreated control on September 12 (Table 56). All fungicide treatments increased yield over the nontreated control except Aproach Prima and Priaxor Xemium. Both were not significantly different from Headline AMP, and Priaxor Xemium was not significantly different from Lucento.

TABLE 55. ***Effect of Fungicide on Foliar Disease Severity***

TREATMENT AND RATE/ACRE[z]	FLS[y] % UC AUGUST 27	FLS[y] % LC AUGUST 27	SBS[y] % LC AUGUST 27	FLS[y] % UC SEPTEMBER 12	FLS[y] % LC SEPTEMBER 12	CLB[y] % SEPTEMBER 12
Nontreated control	1.1 a	0.6 a	1.1 a	10.0 a	1.8 a	22.5 a
Miravis Neo 2.5 SE 13.7 fl oz	0.1 b	0.0 b	0.1 b	1.1 b	0.6 b	11.3 bc
Miravis Top 1.67 SC 13.7 fl oz	0.0 b	0.0 b	0.1 b	1.3 b	0.3 b	11.3 bc
Delaro 325 SC 8.0 fl oz	0.2 b	0.0 b	0.3 b	2.3 b	0.3 b	12.5 bc
Lucento 4.17 SC 5.0 fl oz	0.1 b	0.1 b	0.1 b	0.5 b	0.3 b	16.3 ab
Aproach Prima 2.34 SC 6.8 fl oz	0.1 b	0.0 b	0.4 b	2.5 b	0.3 b	16.3 ab
Priaxor Xemium 4.0 fl oz	0.2 b	0.0 b	0.3 b	2.3 b	0.8 b	16.3 ab
Headline AMP 1.68 SC 10.0 fl oz	0.2 b	0.0 b	0.1 b	2.3 b	0.8 b	12.5 bc
Veltyma 3.34 SC 7.0 fl oz	0.1 b	0.0 b	0.1 b	1.0 b	0.3 b	7.5 c
P-value[x]	0.0002	0.0035	0.0505	0.0001	0.0034	0.0051

[z] Fungicide treatments were applied on August 1 at the beginning pod (R3) growth stage, and all treatments contained a nonionic surfactant (Preference) at a rate of 0.25% v/v.

[y] Foliar disease incidence was rated on a scale of 0–100% of plants within a plot with disease symptoms in the upper canopy (UC) or lower canopy (LC). FLS = frogeye leaf spot, SBS = Septoria brown spot, CLB = Cercospora leaf blight.

[x] Means followed by the same letter are not significantly different based on Fisher's Least Significant Difference test (LSD; α=0.05).

TABLE 56. ***Effect of Fungicide on Foliar Disease Severity and Soybean Yield***

TREATMENT AND RATE/ACRE[z]	SBS[y] % UC	SBS[y] % LC	DEFOLIATION[x] %	HARVEST MOISTURE %	TEST WEIGHT LB/BU	YIELD[w] BU/ACRE
Nontreated control	3.5 a	28.8 a	5.0 a	13.4	55.1	66.7 d
Miravis Neo 2.5 SE 13.7 fl oz	0.0 b	1.0 b	1.0 b	13.4	54.8	76.0 a
Miravis Top 1.67 SC 13.7 fl oz	0.3 b	0.5 b	1.9 b	13.4	54.8	76.8 a
Delaro 325 SC 8.0 fl oz	1.0 b	2.8 b	1.0 b	13.2	54.9	73.3 ab
Lucento 4.17 SC 5.0 fl oz	0.3 b	1.0 b	0.9 b	13.3	54.8	76.0 a
Aproach Prima 2.34 SC 6.8 fl oz	0.0 b	3.8 b	1.0 b	13.3	54.9	68.1 cd
Priaxor Xemium 4.0 fl oz	0.0 b	2.8 b	0.8 b	13.4	55.1	69.8 bcd
Headline AMP 1.68 SC 10.0 fl oz	0.5 b	3.3 b	1.1 b	13.3	54.7	72.0 abc
Veltyma 3.34 SC 7.0 fl oz	0.0 b	0.8 b	0.9 b	13.8	54.8	76.6 a
P-value[v]	*0.0001*	*0.0001*	*0.0001*	*0.1685*	*0.7909*	*0.0007*

[z] Fungicide treatments were applied on August 1 at the beginning pod (R3) growth stage, and all treatments contained a nonionic surfactant (Preference) at a rate of 0.25% v/v.

[y] Foliar disease incidence was rated on a scale of 0–100% of plants within a plot with disease symptoms in the upper canopy (UC) or lower canopy (LC). SBS = Septoria brown spot.

[x] Defoliation visually rated as percentage of leaf loss in plot.

[w] Yields were adjusted to 13% moisture at harvest on October 8.

[v] Means followed by the same letter are not significantly different based on Fisher's Least Significant Difference test (LSD; α=0.05).

FUSARIUM HEAD BLIGHT (FHB) INTEGRATED FUNGICIDE TRIALS IN WHEAT IN SOUTHWESTERN INDIANA, 2019 (WHT19-02.SWPAC)

C. Rocca da Silva, S. Shim, J. D. Ravellette, and D. E. P. Telenko, Department of Botany and Plant Pathology, Purdue University West Lafayette, Indiana 47907-2054

WHEAT (*TRITICUM AESTIVUM* P25R40)

Fusarium head blight, *Fusarium graminearum*
Leaf rust, *Puccinia triticina*

A trial was established at the Southwest Purdue Agricultural Center (SWPAC) in Knox Count, IN. The experiment was a randomized complete block design with four replications. Plots were 7.5 feet wide and 20 feet long and consisted of 12 rows spaced 7.5 inches apart, and the center of each plot was used for evaluation. Standard practices for nonirrigated wheat production in Indiana were followed. On October 8, 2019, wheat cultivar P25R40 was drilled at 7.5-inch spacing. Fungicide applications were applied with a CO_2 pressurized backpack sprayer using a handheld boom fitted with a pair of TJ8001VS nozzles spaced 20 inched apart and directed forward and backward at a 45 degree angle, which delivered 10 gal/acre at 40 psi. Fungicides were applied on May 11 at Feekes growth stage 10.5.1 and on May 16 at Feekes growth stage 10.5.3 (10.5.3 + 4 days). Fusarium head blight (FHB) disease incidence (DI) was measured as the number of infected heads out of 100 in each plot and calculated as a percentage. FHB disease severity (DS) was rated by visually assessing the percentage of those infected heads. FHB index was calculated as (FHB % DI/Average FHB % DS)/100 per plot. DS severity of leaf blotch was rated by visually assessing the percentage of symptomatic tissue on five flag leaves per plot for leaf blotch and five heads per plot for glume blotch. Values for each plot were averaged before analysis. The eight center rows of each plot were harvested with a Kincaid plot combine on June 28, and yields were adjusted to 13.5% moisture. Data were subjected to analysis of variance (SAS 9.4, 2019), and means were compared using Fisher's Least Significant Difference test (LSD; α=0.05).

In 2019, weather conditions were favorable for FHB and leaf rust. FHB was the most prominent disease in the trial. All fungicide programs significantly reduced FHB DI on July 3 (Table 57). Miravis Ace applied at 10.5.4, Miravis Ace followed by Prosaro and Miravis Ace followed by Caramba significantly reduced FHB DS. All fungicide programs reduced FHB Index over the nontreated control except Caramba. No differences were detected between treatments for leaf rust. Deoxynivalenol was reduced by all fungicide programs over the nontreated control (2.34 ppm) except Prosaro, and Miravis Ace followed by Caramba (Table 58). The percentage of Fusarium damaged kernels was significantly reduced with Miravis Ace followed by Prosaro, Miravis Ace followed by Caramba, and Miravis Ace applied at 10.5.4 and 10.5.1 as compared to the nontreated control. All fungicide treatments significantly increased moisture and test weight over the nontreated control. All fungicide programs increased yield over the nontreated control except Caramba.

TABLE 57. ***Effect of Fungicide on Fusarium Head Blight (FHB) and Foliar Diseases in Wheat***

TREATMENT, RATE/ACRE, AND APPLICATION TIMING[z]	FHB % DI[y]	FHB % DS[y]	FHB INDEX[x]	LEAF RUST[w] %
Nontreated control	62.5 a	42.3 a	27.5 a	1.3
Prosaro 6.5 fl oz at 10.5.1	35.3 b	33.6 abc	12.1 b	0.1
Caramba 13.5 fl oz at 10.5.1	46.3 b	35.1 ab	18.6 ab	0.3
Miravis Ace 5.2SC 13.7 fl oz at 10.5.1	36.0 b	35.3 ab	12.6 b	0.4
Miravis Ace 5.2SC 13.7 fl oz at 10.5.4	42.3 b	27.4 bc	11.8 b	0.5
Miravis Ace 5.2SC 13.7 fl oz at 10.5.1 fb Prosaro 6.5 fl oz at 10.5.4	33.3 b	23.5 c	9.2 b	0.2
Miravis Ace 5.2SC 13.7 fl oz at 10.5.1 fb Caramba 13.5 fl oz at 10.5.4	37.5 b	25.6 bc	9.5 b	0.0
P-value[v]	*0.0181*	*0.034*	*0.0286*	*0.2988*

[z] All treatments contained a nonionic surfactant (Preference) at a rate of 0.125% v/v. Plots were inoculated with *Fusarium graminearum* spore suspension (40,000–100,000 spores/ml) 24–26 hours after the treatment at Feekes 10.5.1. Spore suspension was applied at 300 ml/plot with a handheld sprayer on May 31. fb = followed by.

[y] Fusarium head blight (FHB) disease incidence (DI) was measured as the number of infected heads out of 100 in each plot and calculated as a percentage. FHB disease severity (DS) was rated by visually assessing the percentage of infected heads on July 3.

[x] FHB index was calculated as (FHB % DI/Average FHB % DS)/100 per plot.

[w] Disease severity of leaf rust was rated by visually assessing the percentage of symptomatic tissue on five flag leaves per plot on July 3.

[v] Means followed by the same letter are not significantly different based on Fisher's Least Significant Difference test (LSD; α=0.05).

TABLE 58. ***Effect of Fungicide on Deoxynivalenol (DON), Fusarium Damaged Kernels (FDK), and Wheat Yield***

TREATMENT, RATE/ACRE, AND APPLICATION TIMING[z]	DON[y] PPM	FDK[x] %	HARVEST MOISTURE %	TEST WEIGHT LB/BU	YIELD[w] BU/ACRE
Nontreated control	2.4 a	20.0 a	12.8 d	53.6 d	87.3 d
Prosaro 6.5.0 fl oz at 10.5.1	1.8 ab	20.0 a	13.2 bc	54.8 c	93.9 c
Caramba 13.5.0 fl oz at 10.5.1	1.3 cd	15.3 ab	13.0 cd	54.8 c	92.9 cd
Miravis Ace 5.2SC 13.7 fl oz at 10.5.1	1.2 bc	8.8 bc	13.5 a	56.1 a	104.2 a
Miravis Ace 5.2SC 13.7 fl oz at 10.5.4	1.4 bc	7.3 bc	13.5 ab	55.9 ab	98.7 abc
Miravis Ace 5.2SC 13.7 fl oz at 10.5.1 fb Prosaro 6.5.0 fl oz at 10.5.4	0.8 bc	5.3 c	13.4 ab	55.9 ab	102.0 ab
Miravis Ace 5.2SC 13.7 fl oz at 10.5.1 fb Caramba 13.5.0 fl oz at 10.5.4	1.6 abc	10.3 bc	13.5 a	55.3 bc	97.7 bc
P-value[v]	*0.0481*	*0.0091*	*0.0001*	*0.0001*	*0.0003*

[z] All treatments contained a nonionic surfactant (Preference) at a rate of 0.125% v/v. Plots were inoculated with *Fusarium graminearum* spore suspension (40,000–100,000 spores/ml) 24–26 hours after the treatment at Feekes 10.5.1. Spore suspension was applied at 300 ml/plot with a handheld sprayer on May 31. fb = followed by.

[y] Analysisof the mycotoxin deoxynivalenol (DON) completed by the University of Minnesota DON Testing Lab.

[x] FDK = percentage of Fusarium damaged kernels of subsample taken from each plot.

[w] Yields were adjusted to 13.5% moisture at harvest on June 28.

[v] Means followed by the same letter are not significantly different based on Fisher's Least Significant Difference test (LSD; α=0.05).

DAVIS PURDUE AGRICULTURAL CENTER (DPAC)

FIELD-SCALE FUNGICIDE TIMING COMPARISON FOR FOLIAR DISEASES ON CORN IN CENTRAL INDIANA, 2019 (COR19-09.DPAC)

C. Rocca da Silva, S. Shim, J. D. Ravellette, and D. E. P. Telenko, Department of Botany and Plant Pathology, Purdue University West Lafayette, Indiana 47907-2054

CORN (*ZEA MAYS* P0157AM)

A trial was established at the Davis Purdue Agricultural Center (DPAC) in Randolph County, IN. The experiment was a randomized complete block design with four replications. Plots were 30 feet wide and 450 feet long and consisted of 12 rows. The previous crop was soybean. Standard practices for nonirrigated soybean production in Indiana were followed. Corn hybrid P1057AM was planted in 30-inch row spacing at a rate of 30,000 seeds/acre on June 6. All fungicide applications were applied at 20 gal/acre and 40 psi using an Apache 720 sprayer. Fungicides were applied on July 31 at the V9 and on August 12 at the tassel (VT) growth stages. Yields were adjusted to 15.5% moisture. Data were subjected to analysis of variance (SAS 9.4, 2019), and means were compared using Fisher's Least Significant Difference test (LSD; α=0.05).

In 2019, there were no disease detected in the trial. There was no significant treatment effect on percentage of harvest moisture and corn yield (Table 59).

TABLE 59. ***Effect of Fungicide on Moisture and Yield of Corn***

TREATMENT, RATE/ACRE, AND TIMING[z]	HARVEST MOISTURE %	YIELD[y] BU/ACRE
Nontreated control	18.0	135.5
Delaro 325 SC 8.0 fl oz at V9	18.1	143.8
Delaro 325 SC 8.0 fl oz at V9	18.0	140.2
Delaro 325 SC 8.0 fl oz at VT	18.1	135.8
P-value[x]	*0.5914*	*0.3626*

[z] Fungicide treatments were applied on July 31 at the V9 and August 12 at the tassel (VT) growth stages, and all treatments contained a nonionic surfactant (Preference) at a rate of 0.25% v/v.

[y] Yields were adjusted to 15.5% moisture at harvest on November 9.

[x] Means followed by the same letter are not significantly different based on Fisher's Least Significant Difference test (LSD; α=0.05).

FIELD-SCALE FUNGICIDE TIMING COMPARISON FOR FOLIAR DISEASES ON SOYBEAN IN CENTRAL INDIANA, 2019 (SOY19-10.DPAC)

C. Rocca da Silva, S. Shim, J. D. Ravellette, and D. E. P. Telenko, Department of Botany and Plant Pathology, Purdue University West Lafayette, Indiana 47907-2054

SOYBEAN (*GLYCINE MAX* P35T75X)

Frogeye leaf spot, *Cercospora sojina*
Septoria brown spot, *Septoria glycines*

A trial was established at the Davis Purdue Agricultural Center (DPAC) in Randolph County, IN. The experiment was a randomized complete block design with four replications. Plots were 10 feet wide and 500 feet long and consisted of 12 rows. The previous crop was corn. Standard practices for nonirrigated soybean production in Indiana were followed. Soybean variety P34A13X was planted in 7.5-inch row spacing at a rate of 200,000 seeds/acre on June 14 and replanted on June 27. All fungicide applications were applied at 20 gal/acre and 40 psi using an Apache 720 sprayer. Fungicides were applied on July 30 at the V4, August 28 at the beginning pod (R3), and September 9 at the beginning seed (R5) growth stage. Disease ratings were assessed on September 25 at the full seed (R6) growth stage. Frogeye leaf spot and Cercospora leaf blight were rated for disease severity by visually assessing the percentage of symptomatic leaf area in the upper and lower canopies, respectively. The two center rows were harvested on November 4, and yields were adjusted to 13% moisture. Data were subjected to analysis of variance (SAS 9.4, 2019), and means were compared using Fisher's Least Significant Difference test (LSD; α=0.05).

In 2019, frogeye leaf spot and Septoria brown spot were the most prominent diseases but only reached low severity. There was no significant difference between fungicide treatments and the nontreated control for all disease ratings on September 25 (Table 60). There was no significant treatment effect on percentage of harvest moisture and soybean yield.

TABLE 60. ***Effect of Fungicide on Foliar Disease, Moisture, and Yield of Soybean***

TREATMENT, RATE/ACRE, AND TIMING[z]	FLS[y] % UC	FLS[y] % LC	SBS[y] % LC	HARVEST MOISTURE %	D[x] BU/ACRE
Nontreated control	0.9	1.1	1.6	14.4	56.7
Delaro 325 SC 12.0 fl oz at V4	1.5	0.9	0.8	14.6	52.8
Delaro 325 SC 12.0 fl oz at R3	0.8	0.9	1.1	14.4	56.8
Delaro 325 SC 12.0 fl oz at R5	0.6	0.4	1.4	14.3	55.3
P-value[w]	*0.2216*	*0.6831*	*0.2148*	0.002	*0.6816*

[z] Fungicide treatments were applied on July 30 at the V4, August 28 at the beginning pod (R3), and September 9 at the beginning seed (R5) growth stage, and all treatments contained a nonionic surfactant (Preference) at a rate of 0.25% v/v.

[y] Foliar disease incidence was rated on a scale of 0–100% of plants within a plot with disease symptoms. FLS = frogeye leaf spot, SBS = Septoria brown spot.

[x] Yields were adjusted to 13% moisture at harvest on November 4.

[w] Means followed by the same letter are not significantly different based on Fisher's Least Significant Difference test (LSD; α=0.05).

NORTHEAST PURDUE AGRICULTURAL CENTER (NEPAC)

FIELD-SCALE FUNGICIDE TIMING COMPARISON FOR FOLIAR DISEASES ON CORN IN NORTHERN INDIANA, 2019 (COR19-10.NEPAC)

S. Shim, J. D. Ravellette, and D. E. P. Telenko, Department of Botany and Plant Pathology, Purdue University
West Lafayette, Indiana 47907-2054

CORN (*ZEA MAYS* P0157AM)

Gray leaf spot, *Cercospora zeae-maydis*
Northern corn leaf spot, *Bipolaris zeicola*
Southern rust, *Puccinia polysora*
Tar spot, *Phyllachora maydis*

A trial was established at the Northeast Purdue Agricultural Center (NEPAC) in Whitley County, IN. The experiment was a randomized complete block design with four replications. Plots were 10 feet wide and 400 feet long and consisted of four rows, and the two center rows were used for evaluation. The previous crop was soybean. Standard practices for nonirrigated grain corn production in Indiana were followed. Corn hybrid P0157AM was planted in 30-inch row spacing at a rate of 34,000 seeds/acre on June 5. Fungicide treatments were applied on July 9 at the V6, July 24 at the V10, and August 9 at the tassel (VT) growth stages. Disease ratings were assessed on September 13 at the dent (R5) growth stage. Southern rust, northern corn leaf spot, and gray leaf spot were rated for disease severity by visually assessing the percentage of symptomatic leaf area on the ear leaf. Tar spot was rated by visually assessing the percentage of stroma per leaf on five plants in each plot at the ear leaf. The two center rows were harvested on November 5, and yields were adjusted to 15.5% moisture. Data were subjected to analysis of variance (SAS 9.4, 2019), and means were compared using Fisher's Least Significant Difference test (LSD; α=0.05).

In 2019, very little disease developed in plots. Tar spot, southern rust, northern corn leaf spot, and gray leaf spot were the most prominent diseases but only reached low severity. There were no significant fungicide timing effects over the nontreated control for all disease ratings on September 13 (Table 61). There was no significant effect of fungicide timing on moisture and yield.

TABLE 61. ***Effect of Fungicide on Foliar Disease Severity and Corn Yield***

TREATMENT, RATE/ACRE, AND TIMING[z]	GLS[y] %	NCLS[y] %	SR[y] %	TAR SPOT[x] %	HARVEST MOISTURE %	YIELD[w] BU/ACRE
Nontreated control	2.3	1.5	0.8	0.1	20.5	220.9
Headline AMP 1.68 SC 10.0 fl oz at V6	2.2	1.1	0.6	0.1	20.8	225.8
Headline AMP 1.68 SC 10.0 fl oz at V10	1.9	0.9	0.6	0.0	20.9	223.0
Headline AMP 1.68 SC 10.0 fl oz at VT/R1	2.6	0.9	1.0	0.0	20.6	219.0
P-value[v]	*0.9848*	*0.7067*	*0.8802*	*0.8860*	*0.8213*	*0.7197*

[z] Fungicide treatments were applied on July 9 at the V6, July 24 at the V10, and August 9 at the tassel (VT) growth stages, and all treatments contained a nonionic surfactant (Preference) at a rate of 0.25% v/v.

[y] Disease severity was visually assessed as a percentage (0–100%) of symptomatic leaf area on leaf. GLS = gray leaf spot, NCLS = northern corn leaf spot, SR = southern rust.

[x] Tar spot stroma was visually assessed as a percentage (0–100%) of leaf area.

[w] Yields were adjusted to 15.5% moisture at harvest on November 5.

[v] Means followed by the same letter are not significantly different based on Fisher's Least Significant Difference test (LSD; α=0.05).

FIELD-SCALE FUNGICIDE TIMING COMPARISON FOR FOLIAR DISEASES ON SOYBEAN IN NORTHERN INDIANA, 2019 (SOY19-12.NEPAC)

C. Rocco da Silva, S. Shim, J. D. Ravellette, and D. E. P. Telenko, Department of Botany and Plant Pathology, Purdue University West Lafayette, Indiana 47907-2054

SOYBEAN (*GLYCINE MAX* P34A13X)

Frogeye leaf spot, *Cercospora sojina*
Septoria brown spot, *Septoria glycines*
White mold, *Sclerotinia sclerotiorum*

A trial was established at the Northeast Purdue Agricultural Center (NEPAC) in Whitley County, IN. The experiment was a randomized complete block design with four replications. Plots were 10 feet wide and 365 feet long and consisted of four rows, and the two center rows were used for evaluation. The previous crop was corn. Standard practices for nonirrigated soybean production in Indiana were followed. Soybean variety P34A13X was planted in 15-inch row spacing at a rate of 140,000 seeds/acre on June 5. Fungicides were applied on July 10 at V4, August 6 at beginning pod (R3), and August 29 at beginning seed (R5) growth stages. Disease ratings were assessed on August 13 at thc R5 growth stage. Frogeye leaf spot and Septoria brown spot were rated for disease severity by visually assessing the percentage of symptomatic leaf area in the upper and lower canopies, respectively. The two center rows were harvested on October 17, and yields were adjusted to 13% moisture. Data were subjected to analysis of variance (SAS 9.4, 2019), and means were compared using Fisher's Least Significant Difference test (LSD; α=0.05).

In 2019, very little disease developed in plots. Frogeye leaf spot, Septoria brown spot, and white mold were the most prominent diseases but only reached low severity. There was no significant difference between fungicide treatments and the nontreated control for all disease ratings on August 28 (Table 62). There was no significant effect of treatment on soybean yield.

TABLE 62. ***Effect of Fungicide on Foliar Disease and Soybean Yield***

TREATMENT, RATE /ACRE, AND TIMING[z]	FLS[Y] % UPPER CANOPY AUGUST 28	FLS[Y] % LOWER CANOPY SEPTEMBER 20	SBS[Y] % LOWER CANOPY SEPTEMBER 20	WHITE MOLD[X] #/PLOT	YIELD[W] BU/ACRE
Nontreated control	3.0	1.8	3.9	2.4	68.9
Priaxor 4.0 fl oz at V4	1.4	1.0	3.8	2.0	69.8
Priaxor 4.0 fl oz at R3	2.4	1.6	3.6	0.8	71.7
Priaxor 4.0 fl oz at R5	1.3	0.4	3.0	1.1	70.2
P-value[v]	*0.2952*	*0.3612*	*0.9202*	*0.2879*	*0.8538*

[z] Fungicide treatments were applied on July 10 at forth-leaf (V4), August 6 at beginning pod (R3), and August 29 at beginning seed (R5) growth stages, and all treatments contained a nonionic surfactant (Preference) at a rate of 0.25% v/v.

[y] Foliar disease incidence was rated on a scale of 0–100% of plants within a plot with disease symptoms. FLS = frogeye leaf spot, SBS = Septoria brown spot.

[x] White mold disease was assessed by counting the number of plants per plot with symptoms on October 17.

[w] Yields were adjusted to 13% moisture at harvest on October 17.

[v] Means followed by the same letter are not significantly different based on Fisher's Least Significant Difference test (LSD; α=0.05).

SOUTHEAST PURDUE AGRICULTURAL CENTER (SEPAC)

FIELD-SCALE FUNGICIDE TIMING COMPARISON FOR FOLIAR DISEASES ON SOYBEAN IN SOUTHEASTERN INDIANA, 2019 (SOY19-11.SEPAC)

C. Rocco da Silva, S. Shim, J. D. Ravellette, and D. E. P. Telenko, Department of Botany and Plant Pathology, Purdue University West Lafayette, Indiana 47907-2054

SOYBEAN (*GLYCINE MAX* P34A13X)

Frogeye leaf spot, *Cercospora sojina*
Cercospora leaf blight, *Cercospora kikuchii*
Septoria brown spot, *Septoria glycines*

A trial was established at the Southeast Purdue Agricultural Center (SEPAC) in Jennings County, IN. The experiment was a randomized complete block design with four replications. Plots were 10 feet wide and 365 feet long and consisted of four rows, and the two center rows were used for evaluation. The previous crop was corn. Standard practices for nonirrigated soybean production in Indiana were followed. Soybean variety P34A13X was planted in 15-inch row spacing at a rate of 150,000 seeds/acre on July 1. All fungicide applications were applied at 20 gal/acre and 40 psi using Apache 720 sprayer with Trimble CFX monitor for rate and section control and real-time kinematic (RTK) guidance. Fungicides were applied on August 1 at the V4 and August 23 at the beginning pod (R3) growth stages and on September 16 at the beginning seed (R5) growth stage. Disease ratings were assessed on September 25 at the full seed (R6) growth stage. Frogeye leaf spot, Cercospora leaf blight, and Septoria brown spot were rated for disease severity by visually assessing the percentage of symptomatic leaf area in the upper and lower canopies. The two center rows were harvested on October 23, and yields were adjusted to 13% moisture. Data were subjected to analysis of variance (SAS 9.4, 2019) and means were compared using Fisher's Least Significant Difference test (LSD; α=0.05).

In 2019, frogeye leaf spot, Cercospora leaf blight, and Septoria brown spot were the most prominent diseases in the trial and reached low severity. Lucento in all timings reduced frogeye leaf spot in the upper canopy (Table 63). There was no significant difference between treatments and the nontreated control for frogeye leaf spot in the lower canopy, Cercospora leaf blight, and Septoria brown spot. All application timings (V4, R3, and R5) of Lucento increased soybean yield over the nontreated control.

TABLE 63. ***Effect of Fungicide on Foliar Disease and Soybean Yield***

TREATMENT, RATE/ACRE, AND TIMING[Z]	FLS[Y] % UPPER CANOPY	FLS[Y] % LOWER CANOPY	CLB[Y] % UPPER CANOPY	SBS[Y] % LOWER CANOPY	YIELD[X] BU/ACRE
Nontreated control	5.3 a	1.8	6.5	3.0	52.7 b
Lucento 4.17 SC 5.0 fl oz at V4	2.3 b	1.0	8.8	1.4	58.5 a
Lucento 4.17 SC 5.0 fl oz at R3	1.0 b	1.1	6.3	1.3	60.8 a
Lucento 4.17 SC 5.0 fl oz at R5	1.8 b	0.8	8.8	1.5	57.1 a
P-value[w]	*0.0021*	*0.1174*	*0.8444*	*0.1191*	*0.0053*

[z] Fungicide treatments were applied August 1 at the V4, August 28 at the beginning pod (R3), and September 16 at the beginning seed (R5) growth stages, and all treatments contained a nonionic surfactant (Preference) at a rate of 0.25% v/v.

[y] Foliar disease incidence was rated on a scale of 0–100% of plants within a plot with disease symptoms on September 25. FLS = frogeye leaf spot, CLB = Cercospora leaf blight, SBS = Septoria brown spot.

[x] Yields were adjusted to 13% moisture at harvest on October 23.

[w] Means followed by the same letter are not significantly different based on Fisher's Least Significant Difference test (LSD; α=0.05).

APPENDIX: WEATHER DATA

TABLE 64. *Average Monthly Conditions at the Purdue Agronomy Center for Research and Education (ACRE), the Pinney Purdue Agricultural Center (PPAC), and the Southwest Purdue Agricultural Center (SWPAC) in Indiana, 2019*

MONTH	ACRE			PPAC			SWPAC		
	TEMP. MIN[x]. °F	TEMP. MAX[x]. °F	TOTAL PRECIPIT[y]. (IN)	TEMP. MIN[x]. °F	TEMP. MAX[x]. °F	TOTAL PRECIPIT[y]. (IN)	TEMP. MIN[x]. °F	TEMP. MAX[x]. °F	TOTAL PRECIPIT[y]. (IN)
January	18.05	34.43	5.03	15.76	31.46	5.03	23.99	40.74	4.96
February	19.18	36.80	5.35	17.09	33.71	5.35	25.52	43.33	5.42
March	31.29	50.97	7.47	28.26	46.94	7.47	36.30	55.83	7.29
April	40.18	63.42	10.41	37.37	60.09	10.41	46.64	67.99	10.15
May	51.98	74.75	11.38	48.69	70.86	11.38	56.72	77.45	10.90
June	61.00	82.55	10.82	58.67	80.22	10.82	65.07	86.40	10.19
July	62.66	84.21	9.52	60.79	82.72	10.00	67.35	88.16	10.00
August	60.70	83.55	9.71	59.15	81.29	9.71	66.00	88.29	10.05
September	54.28	79.30	8.12	52.77	76.99	8.12	59.30	83.34	8.23
October	43.25	65.92	8.55	41.85	63.30	8.55	47.99	70.03	8.55
November	33.94	52.52	8.24	32.23	49.45	8.24	37.46	56.54	8.24
December	24.45	39.36	7.49	22.42	36.56	7.49	29.16	44.71	7.25

[z] Data courtesy of the Indiana State Climate Office, https://ag.purdue.edu/indiana-state-climate/. Taken from the Purdue Mesonet stations at the Purdue Agronomy Center for Research and Education (ACRE), the Pinney Purdue Agricultural Center (PPAC), and the Southwest Purdue Agricultural Center (SWPAC).

[y] Average minimum and maximum temperatures for each month.

[x] Total precipitation for each month.

ABOUT THE AUTHOR

DARCY E. P. TELENKO is an associate professor and Extension plant pathologist in the Department of Botany and Plant Pathology at Purdue University. Her interdisciplinary research and Extension program are involved in studying the biology and management of soilborne and foliar pathogens of agronomic crops. Telenko is a native of western New York and received her PhD at North Carolina State University. She has published more than 75 peer-reviewed manuscripts and 300 Extension publications. Telenko was awarded the 2024 Leadership Award from the Purdue University Cooperative Extension Specialist Association.

www.ingramcontent.com/pod-product-compliance
Lightning Source LLC
Chambersburg PA
CBHW080016260726
48782CB00028B/448
* 9 7 8 1 6 2 6 7 1 3 5 6 7 *